Common Chemical Reactions

COMMON CHEMICAL REACTIONS

Edited by:

Dr. Tirth

ARCLER
PRESS

www.arclerpress.com

Common Chemical Reactions

Dr. Tirth

Arcler Press
224 Shoreacres Road
Burlington, ON L7L 2H2
Canada
www.arclerpress.com
Email: orders@arclereducation.com

ISBN: 978-1-77469-844-0 (Hardcover)

ABOUT THE EDITOR

Dr. Tirth is presently working as an assistant professor in the department of Chemistry, Faculty of Sciences, Parul University. He did his M.Sc. & Ph.D. in Chemistry from the Maharaja Sayajirao University, Vadodara. His research interest is in Drug Discovery, Synthetic Organic Chemistry and Natural Products isolation. To add to her credits, he has published 15 articles in various international journals. He is having 4 U.S. & 1 Indian patents so far.

TABLE OF CONTENTS

LIST OF FIGURES

LIST OF ABBREVIATIONS

ACCRI	Aviation Climate Change Research Initiative
Ag^+	Silver Ion
BF_3	Boron Trifluoride
C_3H_8	Propane
C_8H_{18}	Octane
CCS	Carbon Capture and Storage
CCU	CO_2 Capture Utilization
CH_4	Methane
CLC	Chemical Looping Combustion
CO	Carbon Monoxide
CO_2	Carbon Dioxide
DC	Direct Current
DRI	Direct Reduction of Iron
EDC	Education for Democratic Citizenship
EDS	Energy Dispersive X-Ray Spectroscopy
EELS	Electron Energy Loss Spectroscopy
EMF	Electromotive Force
EU	European Union
GHG	Greenhouse Gas
H_2O	Water
H_2SO_4	Sulfuric Acid
HBr	Hydrogen Bromide
HCl	Hydrochloric Acid
HF	Hydrogen Fluoride
HI	Hydrogen Iodide
IGCC	Integrated Gasification Combined Cycle
IR	Infrared
$KMnO_4$	Potassium Permanganate
KOH	Potassium Hydroxide
LEDs	Light-Emitting Diodes
NaCl	Sodium Chloride

NH_3	Ammonia
NH_4NO_3	Ammonium Nitrate
NO_2	Nitrogen Dioxide
NOx	Nitrogen Oxides
NPs	Nanoparticles
O_2	Oxygen
OH^-	Hydroxide Ions
PC	Pulverized Coal
PEM FCs	Polymer Electrolyte Membrane Fuel Cells
PM	Particulate Matter
PPE	Personal Protective Equipment
PV	Photovoltaic
ROS	Reactive Oxygen Species
SDS	Sustainable Development Scenario
SO_2	Sulfur Dioxide
SO_3	Sulfur Trioxide
TEM	Transmission Electron Microscopy
UNEP	United Nations Environment Program
UV	Ultraviolet
VP	Vice President
WHO	World Health Organization

ABSTRACT

When chemical bonds between atoms are created or broken, chemical reactions take place. The reactants are the substances that enter a chemical reaction, and the products are the substances produced at the conclusion of the reaction. For instance, the process by which hydrogen-peroxide is broken down into water and oxygen: $2H_2O_2$ (hydrogen peroxide) $=2H_2O$ (water) $+O_2$ (oxygen).

Our reactant in this case is hydrogen peroxide, which is divided into the products, water, and oxygen. Water (H_2O) and oxygen molecules (O_2) are formed by rearranging the atoms that began as hydrogen peroxide molecules. In the above chemical equation, there are additional numbers: the 2s in front of water and hydrogen peroxide. According to this volume, coefficients are the numbers that tell how many molecules are involved in the reaction. They should be included for the equation to be balanced, which means that each element has the same number of atoms on both sides. Conditions should be adjusted to reflect the law of matter conservation, which mentions that no particles are made or destroyed throughout the duration of a common chemical reaction.

PREFACE

In chemical reactions, old bonds are broken to make new ones. Chemical changes occur when two or additional substances interact with one another. When they receive some form of energy, chemical reactions happen. For adequate contact between the reactant molecules, the supplied energy can be in form of light, heat, and mechanical shaking, or electricity. Rearranging the atoms in a chemical reaction result in the creation of new substances.

In nature, chemical reactions are everywhere. Photosynthesis is the process by which plants produce their own food. In daylight, the reaction occurs on the outer layer of the leaf. Chlorophyll, a green pigment found in leaves, uses CO_2, water, and other nutrients from the soil to aid in food production. Oxygen, which is consumed by humans on a daily basis, is produced during this process.

According to the volume, chemical reactions occur in our bodies as well. For our body to function properly, chemical reactions are necessary. Throughout our lives, these chemical reactions continue occurring in our bodies. For instance, the food and acids undergo a chemical reaction throughout digestion. The food is disintegrated into smaller molecules during digestion. Digestive enzymes that aid in food digestion are released by the mouth's salivary glands. With the assistance of digestive enzymes, digestion continues inside the stomach and small intestine.

Chemical reactions can take many different forms. A chemical reaction is also a direct atom rearrangement. Ammonium cyanate, for instance, = H_4N_2CO (Urea). The molecule has the same formula, but the bonds have been reorganized. Isomers are such substances. An isomerization reaction is the name given to this process. When a molecule is disintegrated into smaller pieces, this is called a decomposition reaction. H_2 and O_2 are produced by passing an electric current through water, for instance. This electrolysis reaction – the separating of atoms by passing electricity is an illustration of decomposition reaction: $2H_2O = 2H_2 + O_2$.

This volume mentions that metals are obtained from their oxides and salts through the decomposition reaction. The body breaks down starch into sugar, and proteins break down into smaller molecules called amino acids. Combination-reaction is the opposite of the reaction of decomposition. H_2 and O_2 can be combined to make water: $2H_2 + O_2 = 2H_2O$.

Whitewashing the walls with lime is another illustration of a combination reaction. Slaked lime (calcium hydroxide) is produced when lime (calcium oxide) is added to water. A thin layer of calcium carbonate forms on the walls as a result of a slow reaction between calcium hydroxide and the air's carbon dioxide (CO_2).

A displacement reaction occurs when one group of atoms in a molecule is moved by another. An illustration is the removal of Cu from a salt by another metal. Reduction, which frequently involves a reaction with hydrogen, is the reverse of oxidation. It involves a molecule, atom, or ion gaining an electron. Because neither oxidation nor reduction can occur independently, they always occur simultaneously. If one atom loses electrons, another atom must gain them. $H_2 + F_2 = 2HF$. There are two half reactions to this reaction. The oxidation process involves: $H_2 = 2H^+ + 2e^-$. Furthermore, the reduction reaction involves $F_2 + 2e^- = 2F^-$. Oxygen is reduced and hydrogen is oxidized during water formation from its elements.

CHAPTER

1

Law of Conservation of Matter

Contents

1.1. INTRODUCTION

In the realm of scientific laws, few principles hold as much significance as the law of conservation of matter. This foundational concept in chemistry states that matter, in any form, is never created nor destroyed; rather, it undergoes transformations or rearrangements during chemical reactions. The law of conservation of matter provides a fundamental framework for understanding and predicting the behavior of matter in a variety of contexts, from simple chemical reactions to complex ecological systems.

This is a concept that applies in many fields, especially the field of chemistry and physics, fluid mechanics as well as mechanics. The concept of conservation of the law of matter states that despite how the matter of a component and element rearrange or align themselves, the mass of an object remains constant throughout time. The mass cannot be produced or eliminated. All things that can be seen in the cosmos are made of matter, which cannot be generated or destroyed (Anderson, 2013) (Figure 1.1).

O_{pre} X O_{post} O_{dpost}

O_{pre} = Pre-test
X = Two weeks of instruction on the rock cycle, rock types, and law of conservation of matter
O_{post} = Post-test
O_{dpost} = Delayed post-test

Figure 1.1. *Understanding of the law of conservation of matter.*

Source: Tremel, S., (2011). Investigating Student Understanding of the Law of Conservation of Matter. https://www.semanticscholar.org/paper/Investigating-Student-Understanding-of-the-Law-of-Tremel/2a365fc5764ae8bbb01cab4362a49fe5268c528e.

This law means that matter does not disappear or appear magically but it transforms from one state to another. Matter can be defined as anything that occupies space and it has weight. Matter appears in three ways namely solid gas and liquid. All these substances occupy space and they have weight in them. All these components can as well change from one form to another. For instance, water can change from a solid state to a gaseous state after being boiled. Solid can also change to liquid forms. For example, if you dissolve sugar in water it changes to liquid form but it does not mean that it is not there. When a substance changes its form the matter that was in that substance is conserved.

Generally, in reactions of the chemicals, the total sum of the mass of products is the same as the mass of the reagents used. For example, when water is heated, it transformsintoo a gaseous state. When the water vapor in the gaseous state is heated it turns back to the state of liquid form. This means that as much as the water has changed to the gaseous state, the matter in both substances does not change. This idea applies to any phase transition when a material switches from a particular state to another while retaining the same quantity of matter, such as freezing, sublimation, melting or deposition (Abdishukurovna, 2021).

In the representation of different reactions of chemicals, using an equation, the law of conservation of matter is considered. For the study and creation of chemical processes, the law of conservation of mass is crucial. Experts can anticipate how many products might be produced if they are aware of the types and amounts of the reactants in a certain reaction. By incorporating the rule of conservation of mass into their laboratory procedures, chemical producers can boost the efficiency of their processes. When balancing equations, we must ensure that each element has the same number of atoms on both sides of the expression.

1.2. IMPORTANCE OF THE LAW OF CONSERVATION OF MATTER

The law of conservation of matter, with its far-reaching implications, holds immense importance not only within the realm of chemistry but also extends its influence across a wide range of scientific disciplines. At its core, this fundamental law states that matter cannot be created or destroyed; it can only be transformed or transferred from one form to another (Anderson, 2013).

In the field of chemistry, the law of conservation of matter forms the cornerstone of stoichiometry, which deals with the quantitative relationships between reactants and products in chemical reactions. By adhering to this law, scientists can accurately predict the amount of substances involved in a reaction, determine the limiting reactant, and calculate the yields of products. It provides a crucial framework for understanding and manipulating chemical reaction, facilitating the development of new materials, medicines, and technologies.

Moving beyond the boundaries of chemistry, the law of conservation of matter plays a pivotal role in physics, particularly in the realm of energy-matter equivalence. This principle famously encapsulated in Einstein's equation $E=mc^2$, establishes a profound connection between energy and matter. It asserts that matter can be converted into energy and vice-versa, with the total amount remaining constant. The law provides the foundation for understanding nuclear reactions, such as those occurring in the Sun or in nuclear power plants, where matter is converted into vast amounts of energy (Abdishukurovna, 2021).

Furthermore, the law of conservation of matter guides our comprehension of ecological processes in biology. Ecosystems, comprising various organisms and their environment, rely on the cycling of matter to maintain their balance and sustain life. From the carbon cycle to the nitrogen cycle, the law ensures that essential elements and compounds are conserved and recycled within ecosystems. It emphasizes the cohesiveness of all living organisms and their dependence on the availability and cycling of matter.

In the context of environmental science and sustainable practices, the application of the law of conservation of matter is of paramount importance. With mounting concerns over pollution, climate change, and resource depletion, understanding how matter behaves and can be conserved is crucial for addressing these global challenges. By recognizing that matter cannot simply vanish or appear out of thin air, scientistscan develop strategies to minimize waste, improve recycling and waste management systems, and promote sustainable consumption and production practices (Abdishukurovna, 2021).

The law of conservation of matter serves as a guiding principle for developing a circular economy, where resources are used efficiently and continuously cycled back into the system. By implementing this concept, we can reduce the strain on natural resources, minimize pollution, and move towards a more sustainable future. It also highlights the importance of considering the long-term consequences of our actions, urging us to make informed choices that prioritize the conservation and responsible use of matter.

The law of conservation of matter holds significant importance across multiple scientific disciplines. It forms the skeleton of stoichiometry, facilitates our understanding of energy-matter equivalence in physics, guides ecological processes in biology, and plays a vital role in addressing global challenges through environmental science and sustainable practices. By recognizing and applying this law, we can better comprehend the intricate interconnectedness of matter and make informed decisions to create a more sustainable and balanced world (Anderson, 2013).

1.3. HISTORICAL BACKGROUND

1.3.1. Early Ideas About the Matter

The concept of matter has intrigued humanity since ancient times. From the Greek philosophers' theories of atoms to the alchemical views of substances, various ideas about matter laid the foundation for modern scientific inquiry.

Lavoisier's Axiom, by the knowledge or idea of important particles, defines a fundamental particle as one that cannot be split into other components, hence making the axiom theory to be validated. By using different equations of mathematics it is easy to demonstrate how this law of conservation energy as well as matter operates. In the law of conservation of matter, set research has made its way into the field of science concerned with the quest for order in the natural world in which we live through general systems theory. Behind the law of conservation of matter the scientific theory is atomic theory. It is a scientific idea that states that matter is made up of small particles referred to as atoms. It was accepted in the 18th and 19th centuries because of its explanation of how different gasses would react on different components. Generally, the whole idea was accepted fully in the 19th century.

In contrast to the law of conservation of energy that states that energy cannot be generated but can be changed to different forms, the law of conservation of matter basically stipulates that the mass total must remain the same in a system that is enclosed hence not allowing energy or matter to come out. In the relationship between the law of conservation of matter and energy, at the beginning of the 19th century, the idea of mass was under serious revision by scientists. Through this process of revision one of the best results attained where from Einsteinwho compared both energy and mass and he realized that they were both convertible in both ways. According to the theory some forms of matter can be generated or either destroyed, but the amount of the energy as well as mass remains constant. In his study, he discovered that the mass of the atoms was relatively smaller compared to other masses of neutrons, protons, and electrons that generally make up the nucleus (Anderson, 2013). Einstein's equation is used to express mass-energy equivalence. The equation is written as $E=mc^2$ whereby M represents the mass and E represents energy; c^2 represents the speed of light squared. In the above equation, the energy and mass are representations of the same physical quantity they can be changed within one another. It states that the speed of the light used here may be used to affect both mass as well as energy. It shows that a little light can be used to change a small mass to a large amount of mass and vice versa. Generally, the law of conservation mass energy states that the total energy of a substance is equivalent to the speed of light multiplied by the mass.

During the fusion of the nuclear, some of its mass might be transformed into enormous quantities of energy which leads to the removal of some mass from the initial substance and hence the nuclear that remains lacks some mass. In most cases, the products formed by the chemical vary in mass due to the conversion of energy as well as mass (Abdishukurovna, 2021).

Energy is unable to be generated or destroyed, and there is a constant amount of energy present in everything, which is equal to the first rule of thermodynamics. Basically, Energy is able to be altered into other forms or moved from one location to a different one, but the overall amount of energy cannot vary. In further explanation, the first law of thermodynamics states that the difference between the task done by the system and the heat produced by the system is equivalent to the change that will occur in the closed system given that there is no transfer of matter into the system in or out. During the processes of matter transfer, the amount of energy in the new system is the same as the internal energy of the two initial systems.

Heat transfer is a type of energy transmission. The natural method of transferring energy into or from a system as opposed to doing effort or moving matter, is called heat transfer. The only way to modify the internal power in a thermal system is to transmit energy as heat.

1.3.2. Pioneers of Conservation Laws

The concept of conservation laws gained prominence during the 17th and 18th centuries. The works of scientists such as Robert Boyle, Joseph Black, and Henry Cavendish played a significant role in formulating the laws of conservation of mass and energy (Anderson, 2013).

Antoine Lavoisier, also called the father of Modern Chemistry, discovered that the mass of substances cannot be generated or destroyed by chemical processes in the year 1789, which led to the formation of the law of conservation of mass. This scientist used the process of combustion to develop the law of conservation of matter. He sealed a glass jar and placed a candle that was burning inside it. After the process, the mass of the contents used as well as the weight of the jar remained the same. By that time most of the scientists were investigating how an oxide is formed, as they thought that after burning the metals, they lost weight.

After the experiment of the Lavoisier, he convinced the scientists that the calx came from the result of the gas combining with the metal which instead of the gas being lost led to an increase in mashie changed the thinking of scientists of combustion as they thought it appeared magically. Throughout this development of the law of conservation, it has helped to help scientists in further investigations of matter and understanding of matter in different studies. Other scientists who helped in the development and advancement of this theory are Joseph Proust and he did this by summarization called Proust law. Through this law, chemical compounds are made of the fixed and established ratio which is based on their masses. For example, CO_2 is made up of one carbon of atom as well as two oxygen atoms, which have the same mass. His theory was supported by various scientists at the time which included Claude Berthollet. He stated that at any ratio elements would mix. John Dalton is also another scientist who supported the idea of Proust's and he said that components are made up by different atoms from different elements. This led to a better understanding of matter (Abdishukurovna, 2021).

With current technology and modern formulation, the law of conservation of, the matter has progressed rather than focusing on mass alone to include other studies such as particles and atoms. Application of different mathematical equations to the reactions of chemicals has been a clear definition of advancement and understanding of conservation of matter which has helped in different predictions of results of the reactions.

The development of the law of matter has been advancing with time, as there are more discoveries about matter being done. This helps to understand matter in a better way, in different environments (Anderson, 2013).

1.4. FORMULATION OF THE LAW OF CONSERVATION OF MATTER

1.4.1. Antoine Lavoisier: The Father of Modern Chemistry

Antoine Lavoisier, a remarkable French chemist of the 18th century, holds a prominent place in the history of science as the "Father of Modern Chemistry." Through his groundbreaking research and meticulous experimental work,

Lavoisier revolutionized the understanding of chemical processes and laid the foundation for the formulation of the law of conservation of matter.

Lavoisier was born in Paris in 1743 and developed a passion for science at an early age. Lavoisier's scientific endeavors were characterized by a combination of rigorous experimentation, systematic observation, and logical deductions.

One of Lavoisier's most significant contribution to the field of chemistry was his recognition that combustion is a process involving the combination of a substance with oxygen. By conducting meticulous experiments on the burning of various substances, Lavoisier observed that the total mass of the reactants before the reaction was equal to the total mass of the products after the reaction. This pivotal finding challenged the prevailing phlogiston theory, which proposed that substances released a mysterious fire-like element during combustion.

Lavoisier's meticulous attention to detail led him to propose a new fundamental principle in chemistry—the law of conservation of matter. He articulated this principle in his renowned work, "Traité Élémentaire de Chimie" (Elementary Treatise of Chemistry), published in 1789. According to Lavoisier, matter cannot be created or destroyed in chemical reactions; it merely undergoes rearrangement and transformation. This revolutionary idea laid the groundwork for modern stoichiometry and provided a solid scientific basis for understanding the behavior of matter during chemical reactions (Abdishukurovna, 2021).

Through his experiments, Lavoisier meticulously weighed the substances involved in reactions, taking into account the gases released or absorbed. He demonstrated that the total mass of all reactants always equals total mass of all products, even in cases where the substances underwent significant changes in form or state. Lavoisier's emphasis on quantitative measurements and meticulous record-keeping set a new standard for scientific inquiry, introducing a level of precision and accuracy previously unseen in the field of chemistry.

The formulation of the law of conservation of matter by Lavoisier not only established a fundamental principle in chemistry but also paved the way for further advancements in the field. This principle had a profound impact on the development of stoichiometry, enabling chemists to quantitatively analyze chemical reactions and predict the amounts of substances involved. It provided a solid foundation for the understanding of chemical reactions and the development of new theories, such as atomic theory and the concept of chemical equations. Lavoisier's work extended beyond the confines of chemistry, influencing other scientific disciplines as well. His emphasis on accurate measurement and systematic experimentation set a precedent for scientific methodology, fostering the growth of empirical sciences. Furthermore, his contributions laid the groundwork for the study of chemical thermodynamics and the concept of energy conservation (Anderson, 2013).

In recognition of his remarkable contributions to the field of chemistry, Lavoisier was honored with numerous accolades and became a prominent figure in the scientific community.

Antoine Lavoisier's meticulous experimental work and insightful deductions established him as the "Father of Modern Chemistry." Through his groundbreaking research, he formulated the law of conservation of matter, which revolutionized the understanding of chemical reactions and set a new standard of precision and accuracy in scientific inquiry. Lavoisier's contributions continue to influence the field of chemistry and serve as a testament to the power of scientific curiosity and rigorous experimentation (Abdishukurovna, 2021).

1.4.2. Lavoisier's Contributions to the Law of Conservation of Matter

Antoine Lavoisier, renowned as the "Father of Modern Chemistry," made significant contributions to the formulation and understanding of the law of conservation of matter. Through his groundbreaking experiments and meticulous observations, Lavoisier challenged prevailing theories, revolutionized the field of chemistry, and established a fundamental principle that remains a cornerstone of chemical knowledge.

Lavoisier's exploration of combustion played a crucial role in his formulation of the law of conservation of matter. By conducting meticulous experiments involving the burning of substances, he carefully measured the mass of the reactants before the reaction and the mass of the products after the reaction. In a series of carefully controlled experiments, Lavoisier observed that regardless of the nature of the substances involved or the complexity of the reaction, the total mass of the reactants always equals the total mass of the products.

These observations led Lavoisier to reject the prevailing phlogiston theory, which proposed that substances released a hypothetical fire-like element called phlogiston during combustion. Instead, he postulated that combustion involved the combination of a substance with oxygen from the air. Lavoisier recognized that the increase in weight observed during combustion was due to the combination of the substance with oxygen, rather than the release of phlogiston.

By meticulously weighing the reactants and products, Lavoisier realized that matter could neither be created nor destroyed during chemical reactions. This groundbreaking insight formed the basis for his formulation of the law of conservation of matter. According to this law, the total mass of matter remains constant before and after a chemical reaction, even though the substances may undergo significant changes in form, state, or composition. This principle holds for all chemical reactions, providing a fundamental understanding of the behavior of matter and the underlying principles governing chemical transformations.

Lavoisier's formulation of the law of conservation of matter marked a turning point in the field of chemistry. It challenged long-standing beliefs and laid the groundwork for the development of stoichiometry, the quantitative study of chemical reactions. Lavoisier's emphasis on accurate measurement and meticulous record-keeping introduced a new level of precision and rigor to the field, transforming chemistry from a qualitative discipline into a quantitative science (Abdishukurovna, 2021).

Moreover, Lavoisier's contributions extended beyond combustion. He applied the principles of the law of conservation of matter to various chemical reactions, including the synthesis of compounds and the decomposition of substances. By meticulously measuring and analyzing the mass changes in these reactions, he further solidified the understanding that matter is conserved, providing evidence for the validity of his law.

Lavoisier's work not only had a profound impact on the scientific community but also had practical implications. His understanding of the law of conservation of matter led to advancements in chemical analysis, improved methods of quantitative measurement, and the development of chemical nomenclature. His work also laid the foundation for the concept of chemical equations, where the balancing of reactants and products is based on the conservation of matter (Anderson, 2013).

Furthermore, Lavoisier's contributions to the law of conservation of matter played a pivotal role in the development of other scientific disciplines. His recognition that matter cannot be created or destroyed influenced the field of thermodynamics and the study of energy conservation. Lavoisier's work provided a crucial link between the conservation of matter and the conservation of energy, shaping our understanding of the interconnectedness of physical and chemical processes. Antoine Lavoisier's experiments and meticulous observations transformed our understanding of chemical reactions and led to the formulation of the law of conservation of matter. His groundbreaking work challenged prevailing theories, establishing that matter is neither created nor destroyed during chemical reactions. Lavoisier's contributions revolutionized the field of chemistry, paving the way for further scientific progress and leaving an enduring legacy in the annals of scientific discovery.

1.4.3. Development of the Law Through Time

The law of conservation of matter, formulated by Antoine Lavoisier, underwent further development and revision in the years following its initial establishment. Notable scientists such as Joseph Proust, John Dalton, and Amadeo Avogadro played instrumental roles in expanding our understanding of the composition and behavior of matter, thereby solidifying the principles of the law.

Joseph Proust, a French chemist of the late 18th century, made significant contributions to the understanding of chemical composition. His experiments were focused on the study of compounds and their elemental composition. Proust discovered that regardless of the source or method of preparation, a compound always contained the same elements in fixed proportions by mass. This observation became known as the law of definite proportions or the law of constant composition. Proust's work provided crucial evidence in support of the law of conservation of matter by demonstrating that matter was conserved not only in reactions but also in the formation of compounds.

John Dalton, an English chemist, and physicist, further expanded upon the law of conservation of matter through his atomic theory. Dalton proposed that matter is composed of indivisible particles called atoms, each with its unique properties. According to his theory, atoms combine in simple, whole-number ratio to form compounds. Dalton's atomic theory provided a conceptual framework for understanding how matter behaves during chemical reactions and reinforced the idea that matter is conserved. By attributing the conservation of matter to the preservation and rearrangement of atoms, Dalton's work solidified the fundamental principles of the law of conservation of matter (Anderson, 2013).

Amadeo Avogadro, an Italian scientist of the 19th century, contributed to the development of the law of conservation of matter by introducing the concept of the mole. Avogadro proposed that equal volumes of gases, at the same temperature and pressure, contain an equal number of particles. This concept, known as Avogadro's law, laid the foundation for understanding the relationship between the mass and the number of particles in a substance. Avogadro's work played a crucial role in establishing a quantitative link between the macroscopic world of measurements and the microscopic world of atoms and molecules. The introduction of the mole as a unit of measurement allowed chemists to accurately determine the amounts of substances involved in reactions and further validated the principle of conservation of matter.

Together, the contributions of Proust, Dalton, and Avogadro refined and expanded upon the law of conservation of matter. Proust's law of definite proportions emphasized the fixed composition of compounds, supporting the idea of matter is conserved during chemical reactions. Dalton's atomic theory provided a framework for understanding how matter is rearranged at the atomic level, reinforcing the concept of conservation. Avogadro's introduction of the mole as a unit of measurement enabled precise quantification of the amounts of substances involved, consolidating the quantitative nature of the law.

Over time, these scientific advancements led to a more comprehensive understanding of the law of conservation of matter. From Lavoisier's initial formulation to the contributions of Proust, Dalton, and Avogadro, the law emerged into a fundamental principle that governs chemical reactions and guides our understanding of the behavior of matter. It became clear that matter is neither created nor destroyed but undergoes transformations and rearrangements, while the total mass remains constant.

The development of the law of conservation of matter continued through the contributions of Joseph Proust, John Dalton, and Amadeo Avogadro. Their work solidified the principles of the law, emphasizing the fixed composition of compounds, the rearrangement of atoms, and the quantitative measurement of matter. Through their collective efforts, our understanding of the law of conservation of matter established, providing a solid foundation for the study of chemistry and its applications in various scientific disciplines (Abdishukurovna, 2021).

1.4.3.1. Experimental Evidence

Chemical reactions have played a pivotal role in solidifying the concept of the law of conservation of matter throughout scientific history. These reactions have been extensively studied and meticulously observed, providing compelling evidence for the fundamental principle that matter cannot be created or destroyed.

Countless experiments have been conducted, with meticulous attention to detail, to quantify the mass of reactants before a chemical reaction takes place and compare it to the mass of products after the reaction completion. These weight measurements have consistently shown that the total mass of substances involved in a reaction remains constant. This finding supports the notion that matter is indestructible and remains conserved during chemical transformations.

Additionally, the pioneering work of Antoine Lavoisier in combustion experiments further solidified the understanding of the law of conservation of matter. Lavoisier meticulously investigated the reaction between oxygen and various substances, including the combustion of fuels and the oxidation of metals. Through his meticulous observations and measurements, Lavoisier demonstrated that the mass of the reactants, such as the fuel and oxygen, equaled the mass of the products, including the resulting gases and ashes. This finding provided compelling evidence that matter is neither created nor destroyed but simply rearranged during chemical reactions (Anderson, 2013).

According to Dalton's atomic theory, the atoms in reaction undergo reconstitution during chemical reactions to form the products, not that new atoms are formed or that some get destroyed, so the matter is neither lost nor gained during chemical reactions, but only changes from one form to another.

The appropriate solutions of additional chemicals can be used to do the aforementioned experiment. Examples include sodium tetraoxosulfate (VI) and calcium trioxonitrate (V), lead trioxonitrate (V) and potassium iodide, barium chloride and dilute tetraoxosulfate (VI) acid, and barium chloride and calcium trioxonitrate (V) and dilute tetraoxosulfate (VI) acid. In 1908, Landolt carried out a more exact experiment to illustrate the law of conservation of matter. The apparatus used for this is given below:

The law of conservation of matter has far-reaching implications in various scientific fields, such as chemistry, physics, and environmental science. It forms the basis for stoichiometry and allowing scientists to predict the quantities of reactants and products in a chemical reaction. Furthermore, it underpins our understanding of the Earth's biogeochemical cycles, where matter is continuously cycled and transformed while adhering to the principle of conservation. The experimental evidence provided by weight measurements, combustion experiments, and countless chemical reactions has irrefutably confirmed the law of conservation of matter. These investigations have consistently demonstrated that matter remains constant, reinforcing the notion that it cannot be created nor destroyed but only undergoes a transformation. The law of conservation of matter stands as a fundamental principle in understanding the behavior of matter and its interactions in the natural world (Abdishukurovna, 2021).

1.5. IMPLICATIONS IN CHEMISTRY

1.5.1. Balancing Chemical Equations

Stoichiometry, the study of the quantitative relationships in chemical reactions, is an essential tool that enables scientists to delve into the intricate details of chemical transformations. At the heart of stoichiometry lies the fundamental principle of the law of conservation of matter, which states that matter cannot be created or destroyed in a chemical reaction. Instead, it undergoes a rearrangement of atoms, preserving the total number of atoms of each element.

The law of conservation of matter plays a pivotal role in stoichiometric calculations which provide a guiding principle for chemists to determine reactant and product quantities accurately. By ensuring that the number of atoms of each element is equal on both sides of a balanced chemical equation, chemists uphold the principle of mass conservation.

Balancing chemical equations is a crucial step in stoichiometry, as it allows chemists to establish a correspondence between the reactants and products involved in a reaction.

This balancing process involves adjusting the coefficients in the chemical equation to confirm the total number of atoms of each element remains constant on both the reactant and product sides. By achieving a balanced equation, chemists uphold the law of conservation of matter and establish a quantitative relationship between the different species involved in the reaction (Anderson, 2013) (Figure 1.2).

Once a chemical equation is balanced, stoichiometric calculations can be performed with confidence. These calculations involve the use of conversion factors derived from the balanced equation, such as molar ratio and stoichiometric coefficients. This ratio allows scientists to relate the amounts of substances involved in a reaction on a mole-to-mole basis, providing a quantitative understanding of the reaction's composition and stoichiometry (Abdishukurovna, 2021).

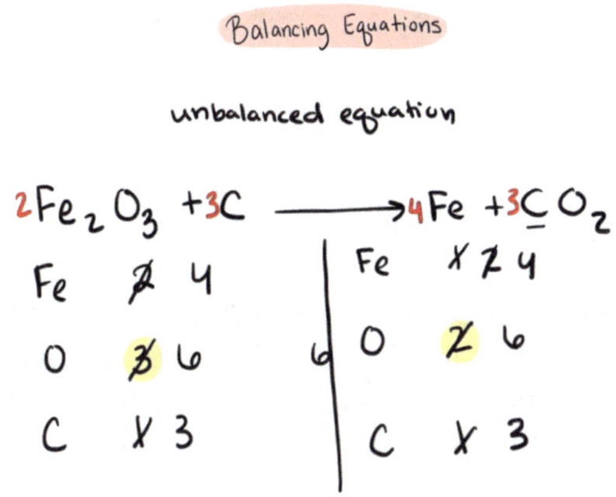

Figure 1.2. *Balancing chemical equation.*

Source: Li, M., Wang, Y., Stone, H., & Turki, N., (2021). Teaching Introductory Chemistry Online: The Application of Socio-Cognitive Theories to Improve Students' Learning Outcomes. Education Sciences. – Scientific Figure on ResearchGate. Available from: https://www.researchgate.net/figure/A-screenshot-of-the-micro-lecture-on-balancing-equations_fig1_349662042.

Through stoichiometric calculations, scientists can determine the quantities of reactants needed to produce a certain amount of product or vice versa. This information is crucial for various applications, such as optimizing reaction conditions, predicting the yield of a reaction, and designing experiments. By understanding the stoichiometry of a reaction, chemists can make informed decisions regarding the amounts of reagents to use, ensuring efficient resource utilization and minimizing waste.

Moreover, stoichiometry helps chemists identify limiting reactants and calculate theoretical yields. The limiting reactant is the substance that is completely consumed in a reaction, limiting the amount of product that can be formed. By quantitatively analyzing the stoichiometry of the reaction, chemists can determine which reactant is limiting and calculate the maximum amount of product that can be obtained, known as the theoretical yield. This information is vital for assessing reaction efficiency, evaluating the economic viability of a process, and optimizing production (Anderson, 2013) (Figure 1.3).

Composition,	Name		Raw material mass (grams)		
x	Sample	Code	CoO	ZnO	Fe_2O_3
0	$CoFe_2O_4$	CZF-0	2.5989	-	7.4011
0.25	$Co_{0.75}Zn_{0.25}Fe_2O_4$	CZF-1	1.9080	0.8473	7.2447
0.5	$Co_{0.5}Zn_{0.5}Fe_2O_4$	CZF-2	1.2457	1.6595	7.0948
0.75	$Co_{0.25}Zn_{0.75}Fe_2O_4$	CZF-3	0.6102	2.4388	6.9510
1	$ZnFe_2O_4$	CZF-4	-	3.1872	6.8128

Figure 1.3. *Stoichiometry calculation sample.*

Source: Johan, A., Adi, W., Arsyad, F., & Setiabudidaya, D., (2019). Analysis of crystal structure of magnetic materials $Co_{1-x}Zn_xFe_2O_4$. Journal of Physics: Conference Series. – Scientific Figure on ResearchGate. Available from: https://www.researchgate.net/figure/Stoichiometry-calculation-sample-Co-1-x-Zn-x-Fe-2-O-4_tbl1_335157104.

In experimental settings, stoichiometry plays a critical role in designing accurate experiments. By understanding the stoichiometry of a reaction, scientists can determine the appropriate amounts of reactants to use, ensuring that the reaction proceeds as intended and minimizing experimental errors. Stoichiometric calculations guide researchers in selecting the optimal stoichiometric ratios and quantities for their experiments, facilitating reproducibility and precise data analysis.

Stoichiometry and the law of conservation of matter are closely intertwined in the realm of chemistry. Stoichiometry relies on the principle of mass conservation to establish quantitative relationships between reactants and products in a chemical reaction. By balancing chemical equations and performing stoichiometric calculations, scientists can determine reactant and product quantities, predict yields, optimize reaction conditions, and design experiments properly. Through the application of stoichiometry, chemists unlock a deeper understanding of the quantitative aspects of chemical reactions and harness this knowledge to drive advancements in various fields of study and practical applications (Abdishukurovna, 2021).

1.6. CONSERVATION OF MATTER IN PHYSICS

1.6.1. Mass-Energy Equivalence

The law of conservation of matter, a fundamental principle in physics, reveals its profound connection to Einstein's mass-energy equivalence principle. This revolutionary concept revolutionized our understanding of the interplay between matter and energy. Einstein's equation, $E=MC^2$ encapsulates the profound relationship between mass and energy, suggesting that they are interchangeable entities.

The mass-energy equivalence principle postulates that mass can be converted into energy and vice versa. It implies that matter possesses an inherent energy content, and this energy can manifest in various forms. Conversely, energy can condense into matter under specific conditions. This principle demonstrates that matter and energy are not distinct entities but two different aspects of the same underlying reality.

Einstein's equation, $E=mc^2$, is the mathematical expression of this profound relationship. In this equation, E represents energy, m represents mass, and c^2 represents the speed of light squared, which is an enormous constant value. The equation implies that a small amount of mass contains an immense amount of energy, and conversely, a large amount of energy can condense into a minute mass. This realization fundamentally altered our perception of the universe and opened up new avenues of exploration.

The implications of mass-energy equivalence fundamentally challenges the traditional notion that matter and energy are distinct and unrelated entities. Instead, it unifies them under a single framework, emphasizing their deep interdependence. This principle has been confirmed by numerous experiments and observations including nuclear reactions, particle physics experiments, and astrophysical phenomena.

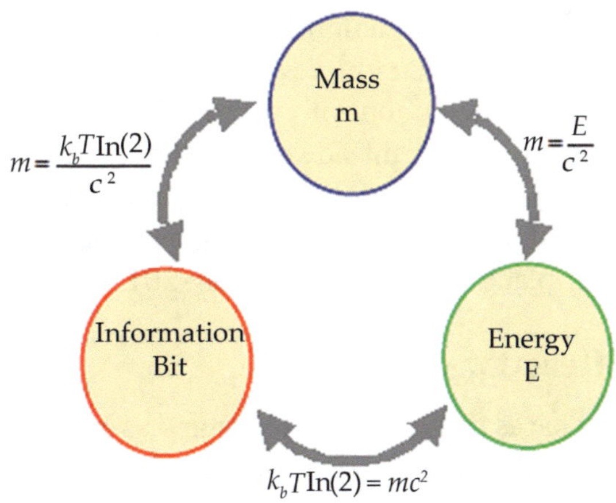

Figure 1.4. *Diagrammatic representation of the mass-energy-information equivalence.*

Source: Vopson, M., (2019). The Information Content of the Universe and the Implications for the Missing Dark Matter. – Scientific Figure on ResearchGate.

Available from: https://www.researchgate.net/figure/Diagrammatic-representation-of-the-mass-energy-information-equivalence_fig1_333751969.

The practical applications of mass-energy equivalence are profound. It forms the basis of nuclear energy, where the conversion of a small amount of matter into energy releases an enormous amount of power. This principle underlies the operation of nuclear reactors and the potential of nuclear fusion as a clean and abundant energy source. The understanding of mass-energy equivalence also plays a pivotal role in particle accelerators, where high-energy collisions can generate new particles by converting energy into matter (Anderson, 2013) (Figure 1.4).

Furthermore, mass-energy equivalence has implications for cosmology and the study of the universe's origins. In the early stages of the universe, when temperatures were extremely high, energy could spontaneously condense into matter, leading to the creation of particles and antiparticles. As the universe expanded and cooled, matter, and antimatter annihilate each other, leaving behind a slight excess of matter that eventually formed galaxies, stars, and ultimately, life as we know it. The understanding of mass-energy equivalence helps unravel the mysteries of the early universe and its evolution process. (Abdishukurovna, 2021).

The law of conservation of matter finds a profound connection to Einstein's mass-energy equivalence principle. This principle demonstrates that matter and energy are not separate entities but intricately linked aspects of the same underlying reality. The equation $E=MC^2$ captures the transformative potential of mass-energy conversion, revealing that even a small amount of mass contains an immense amount of energy. Understanding mass-energy equivalence has led to significant advancements in fields such as energy production, particle physics, and cosmology. By recognizing the deep interplay between matter and energy, we can gain a deeper understanding of the universe and our place within it (Anderson, 2013).

1.6.2. Nuclear Reactions

Nuclear reactions, such as fission and fusion, stand as extraordinary examples that challenge the strict conservation of matter as defined by classical principles. These reactions involve the manipulation of atomic nuclei, where the transformation of matter into energy becomes apparent. To comprehend these intricate processes, an extension of the law of conservation of matter is necessary, one that accounts for the phenomenon of mass deficit and the release of immense amounts of energy.

In nuclear reactions, the nucleus of an atom undergoes a transformative change, leading to the formation of different nuclei and the liberation of energy. These reactions can take two main forms: fission, where a heavy nucleus splits into two or more smaller nuclei, and fusion, where lighter nuclei merge to form a heavier nucleus. Both processes involve a conversion of matter into energy, revealing the profound implications of Einstein's mass-energy equivalence principle.

In nuclear fission, a heavy nucleus, such as uranium-235, is bombarded with neutrons, leading to its fragmentation into smaller nuclei, accompanied by the release of additional neutrons and a significant amount of energy. During this process, a small fraction of the nucleus's mass is converted into energy, according to Einstein's equation, $E=mc^2$. The liberated energy manifests in the form of the kinetic energy of the fragments and the released neutrons, as well as electromagnetic radiation, such as gamma rays. The conversion of mass into energy in fission reactions is responsible for the immense power generation in nuclear reactors and the destructive force of nuclear weapons (Figure 1.5).

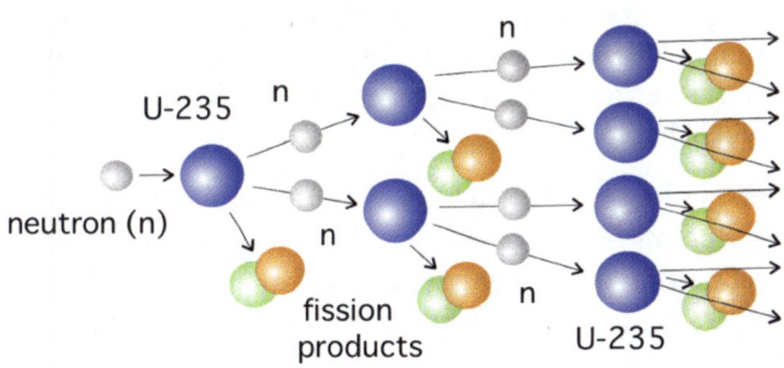

Figure 1.5. *A model of nuclear fission chain reaction.*

Source: Ochiai, E., (2014). Nuclear Power and Radiation. – Scientific Figure on ResearchGate. Available from: https://www.researchgate.net/figure/A-Model-of-Nuclear-Fission-Chain-Reaction-modified-from-Fig-1724-in-Atkins-and_fig1_299687377.

In contrast, nuclear fusion involves the combination of lighter nuclei to form a heavier nucleus. This process is the source of energy in the Sun and other stars. In the core of stars, where temperatures and pressures are extremely high, hydrogen nuclei fuse together to form helium nuclei. During fusion, a fraction of the combined mass is converted into energy and releases a tremendous amount of light and heat. Fusion reactions are highly desirable on Earth as a potentially clean and sustainable energy source. However, achieving controlled fusion reactions remains a significant scientific and technological challenge due to the extreme conditions required for the fusion process to occur (Anderson, 2013) (Figure 1.6).

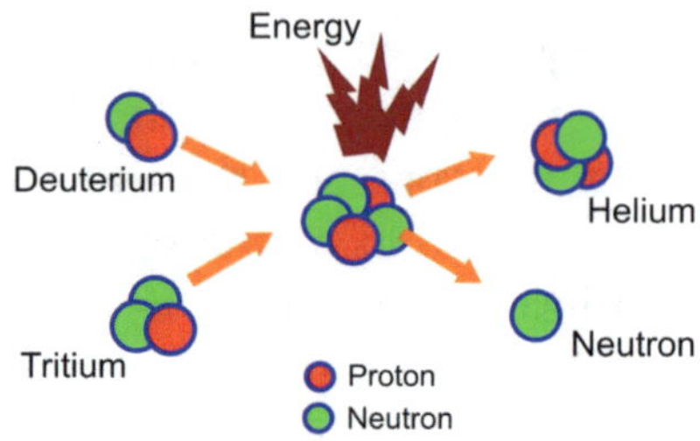

Figure 1.6. *Schematic representation of a nuclear fusion reaction.*

Source: Herndon, J., (2020). Trump administration begins to recognize the loss of scientific integrity: Refuting a political hitjob. Advances in Social Sciences Research Journal. – Scientific Figure on ResearchGate. Available from: https://www.researchgate.net/figure/Schematic-representation-of-a-nuclear-fusion-reaction_fig4_342019575.

The understanding of nuclear reactions necessitates an expansion of the law of conservation of matter to include the concept of mass deficit. While classical conservation laws strictly demand that matter be conserved in all reactions, nuclear reactions showcase that a minute portion of the mass is transformed into energy. This mass deficit is precisely accounted for in Einstein's mass-energy equivalence equation, which demonstrates that even a small change in mass results in an enormous release of energy.

Furthermore, the conservation of other fundamental quantities, such as charge and lepton number, remains intact in nuclear reactions, ensures the overall balance of these properties. While the strict conservation of matter is compromised, the conservation of these other essential properties provides a more comprehensive understanding of the underlying principles governing nuclear processes (Abdishukurovna, 2021).

Nuclear reactions challenge the traditional conservation of matter due to the conversion of mass into energy. Fission and fusion reactions demonstrate that even a small amount of matter can release an enormous amount of energy. To comprehend these processes, an extension of the law of conservation of matter is required, one that incorporates the concept of mass deficit and the remarkable consequences of mass-energy equivalence. The understanding of nuclear reactions and the ability to harness their energy potential have immense implications for power generation, scientific research, and our exploration of the universe.

1.7. CONSERVATION OF MATTER IN BIOLOGY AND ECOLOGY

1.7.1. Biogeochemical Cycles

Biogeochemical cycles play a fundamental role in maintaining the balance of matter in biological systems and the environment, showcasing the remarkable principle of the conservation of matter. These cycles, including the carbon cycle, nitrogen cycle, and water cycle, illustrate the intricate web of interactions between living organisms and their surrounding ecosystems.

The carbon cycle, one of the most crucial biogeochemical cycles, involves the continuous movement of carbon atoms between the atmosphere, living organisms, and the Earth's crust. It begins with the process of photosynthesis, where plants and other photosynthetic organisms utilize CO_2 from the atmosphere to produce organic compounds. Through this process, carbon becomes incorporated into the tissues of plants and forms the foundation of the food chain.

As organisms consume plants or other animals, the carbon stored in their tissues is passed along, allowing energy to flow through the ecosystem. During respiration, carbon is released back into the atmosphere in the form of CO_2, contributing to the greenhouse effect. Additionally, when organisms die or excrete waste, decomposition by bacteria and other decomposers return carbon to the environment, completing the cycle (Anderson, 2013).

Similarly, the nitrogen cycle illustrates the conservation of matter through the continuous circulation of nitrogen in various forms. Nitrogen gas, which makes up about 78% of the Earth's atmosphere, cannot be directly used by most organisms. However, few bacteria have the ability to convert atmospheric nitrogen into ammonia through a process called nitrogen fixation. This ammonia can be converted into nitrate and nitrite ions by other bacteria in the soil.

Plants take up these nitrogen compounds from the soil and incorporate them into their tissues. When animals consume plants or other animals, they acquire nitrogen-containing molecules, which are then used to build their own proteins and nucleic acids. Decomposition of organic matter by decomposers releases nitrogen back into the soil as ammonia, completing the cycle (Abdishukurovna, 2021).

The water cycle, also known as the hydrological cycle, demonstrates the conservation of matter on a global scale. It involves the continuous movement of water between the atmosphere, land, and oceans. Water evaporates from bodies of water and transpires from plants, forming water vapor in the atmosphere. Condensation occurs when the water vapor cools and forms clouds. Eventually, precipitation returns water to the Earth's surface, replenishing bodies of water and infiltrating into the soil.

Through the water cycle, the amount of water on Earth remains relatively constant, with water molecules continuously transitioning between different states and locations. This cycle ensures a steady supply of water for ecosystems, sustaining life and enabling various processes such as photosynthesis and nutrient transport.

Overall, biogeochemical cycles exemplify the law of conservation of matter in biology and ecology. They highlight the interconnectedness of living organisms and their environment, demonstrating the continuous flow and transformation of matter essential for the functioning and stability of ecosystem. By understanding and studying these cycles, scientists, and researchers can gain valuable inputs into the intricate mechanisms that sustain life on Earth.

1.7.2. Ecosystem Dynamics

The law of conservation of matter is a fundamental principle that profoundly influences the study of ecology and ecosystem dynamics. It emphasizes the interconnectedness of organisms within their environment and underscores the crucial role of matter flow and transformation in shaping ecological systems.

Ecosystems are complex networks of living organisms and their surrounding physical environment, where matter continuously moves and cycles through various components. This flow of matter, encompassing nutrients, energy, and other essential elements, directly impacts the distribution and abundance of species, nutrient availability, and overall stability of the ecosystem.

One key aspect influenced by the conservation of matter is the distribution and abundance of species within an ecosystem. Different organisms have specific requirements for matter, including nutrients and energy, which are obtained from their environment. As matter flows through the ecosystem, it becomes available to different organisms through processes such as predation, herbivory, and scavenging.

The availability and distribution of matter play a crucial role in determining which species can thrive in a particular habitat. For example, nutrient-rich area may support a higher diversity of plant species, providing more resources for herbivores and subsequently attracting a greater variety of predators. In contrast, nutrient-poor regions may have limited plant growth, leading to reduced species diversity throughout the food chain (Anderson, 2013).

Moreover, the conservation of matter also influences nutrient availability within ecosystems. Nutrients, such as nitrogen, phosphorus, and potassium, are essential for the growth and development of organisms. These nutrients are cycled through the ecosystem via biogeochemical processes, including decomposition, mineralization, and nutrient uptake by plants.

When matter is returned to the environment through processes like decomposition, it becomes available for uptake by other organisms. This recycling of nutrients ensures their continual availability within the ecosystem, supporting the growth of primary producers and subsequent trophic levels. The conservation of matter emphasizes the importance of maintaining nutrient cycling processes to sustain healthy ecosystems and maintain nutrient balance or limitations (Abdishukurovna, 2021).

Additionally, the conservation of matter contributes to the overall stability of ecosystems. Matter flow and transformation help regulate various ecological processes, including energy transfer, population dynamics, and ecosystem resilience. For example, the flow of matter and energy through food chain and food web allows for the transfer of energy from producers to consumers and decomposers, maintaining energy balance within the system.

Furthermore, matter flow and transformation influence population dynamics within ecosystems. Changes in matter availability, such as fluctuations in nutrient levels or the introduction of new matter sources, can impact the growth and reproduction of organisms. These changes may lead to shifts in population size, interaction, and community structure, ultimately influencing the overall stability and functioning of the ecosystem.

Overall, the conservation of matter is a critical principle that underlies the study of ecosystem dynamics. It emphasizes the continuous flow and transformation of matter within ecological systems, impacting species distribution, nutrient availability, and overall ecosystem stability. Understanding and managing these matter flows is essential for maintaining the health and resilience of ecosystems in the face of environmental changes and human impacts.

1.8. CHALLENGES AND LIMITATIONS

1.8.1. Nuclear Reactions and Mass Deficiency

The concept of conservation of matter, which states that matter cannot be created or destroyed in ordinary chemical reactions, encounters significant challenges when dealing with nuclear reactions. Unlike chemical reactions, nuclear reactions involve changes in the atomic nucleus, leading to the conversion of mass into energy. This phenomenon gives rise to a perplexing issue known as mass deficiency.

When certain elements undergo nuclear reactions, such as nuclear fission or fusion, a minute amount of mass is converted into an enormous amount of energy according to Einstein's famous equation, $E=MC^2$. This mass-energy equivalence principle suggests that mass and energy are interchangeable entities. The conversion of a small amount of mass can yield an enormous amount of energy, as exemplified by the destructive power of nuclear weapons and the immense energy generated in stars (Abdishukurovna, 2021).

The realization that mass can be transformed into energy and vice versa revolutionized our understanding of matter and energy. It shattered the notion that matter is a static entity, separate from energy, and revealed the interconnectedness of these fundamental components of the universe. While the law of conservation of matter seemingly fails to hold nuclear reactions, it is reconciled through the mass-energy equivalence principle. Above principle allows for a more comprehensive understanding of matter's transformations, encompassing both ordinary chemical reactions and extraordinary nuclear reactions (Anderson, 2013).

1.8.2. Extreme Conditions and Fundamental Particles

Beyond the realm of everyday chemical reactions, the law of conservation of matter faces further limitations when confronted with extreme conditions and the interactions of fundamental particles. In these exceptional circumstances, the strict adherence to the conservation of matter may falter due to the influence of quantum effects and the behavior of fundamental particles. At the most fundamental level, the principles of quantum mechanics monitor the behavior of particles and their interactions. According to these principles, particles can exist in multiple states simultaneously, and their properties can be probabilistic in nature. Under extreme conditions, such as those encountered in particle accelerators or the early universe, the energies involved can disrupt the delicate balance of matter conservation.

In particle accelerators, scientists collide particles at high speeds and energies, recreating conditions similar to those present in the early universe. In these collisions, matter and antimatter particles can be produced, and under certain circumstances, they can annihilate each other, resulting in the release of energy. This annihilation process demonstrates that any matter can be created and destroyed under specific scientific frameworks that encompass the behavior of fundamental particles.

Similarly, during the early moments of the universe's existence, incredibly high energies prevailed, enabling the spontaneous creation of matter and antimatter pairs from pure energy. Although the majority of these particles annihilated each other, a slight asymmetry in their numbers allowed a small fraction of matter to persist, eventually forming the basis of the universe as we know it.

It is important to note that these scenarios, involving extreme conditions and fundamental particles, operate within specialized scientific frameworks that transcend the scope of everyday chemical reactions. While the law of conservation of matter forms a fundamental principle in understanding and predicting the behavior of matter in our day-to-day lives, it requires refinement and expansion to account for the extraordinary conditions present in particle physics and cosmology (Abdishukurovna, 2021).

In conclusion, the law of conservation of matter stands as one of the fundamental principle of nature, revealing the intricate balance and interplay of substances within the universe. This law, derived from centuries of scientific observations and experiments, asserts that matter cannot be created nor be destroyed but can only be transformed from one form to another. It emphasizes the inherent stability and persistence of matter, demonstrating that the total mass of a closed system remains constant over time.

Through the lens of the law of conservation of matter, we have gained profound insights into various fields of study, ranging from chemistry to ecology, from physics to environmental science. It serves as a guiding principle in understanding chemical reactions, providing the foundation for stoichiometry and the understanding balanced equations. Additionally, this law has crucial applications in fields such as waste management, pollution control, and sustainability, reminding us of the urgent need to adopt practices that minimize the generation and disposal of waste (Anderson, 2013).

Furthermore, the law of conservation of matter simplifies the interconnectedness of natural systems. It highlights the intricate cycles and exchanges of matter that occur within and between ecosystems, emphasizing the importance of maintaining equilibrium and harmony in our environment. By recognizing and respecting this law, we can strive for a more sustainable future, where resources are conserved, waste is minimized, and the delicate balance of our planet is preserved.

Overall, the law of conservation of matter is a cornerstone principle that guides our understanding of the physical world. It serves as a constant reminder that matter, in all its forms, is a precious and finite resource that must be treated with care and responsibility. By embracing this principle, we can forge a path toward a more sustainable and harmonious coexistence with our natural surroundings (Anderson, 2013).

CHAPTER

2 Writing Chemical Reactions

Contents

2.1. INTRODUCTION

In 1997, Nathan Zohner, a ninth-grader at Eagle Rock Junior High School in Idaho Falls, Idaho, completed a science fair project on the negative effects of DHMO (dihydrogen monoxide). He distributed copies of his report to his classmates and asked them to sign a petition to make the substance illegal, claiming that DHMO is found in cancer patients' tumor cells, that it can cause severe burns when in the vapor state, and that it can be fatal if consumed accidentally (Davis, 2012) (Figure 2.1).

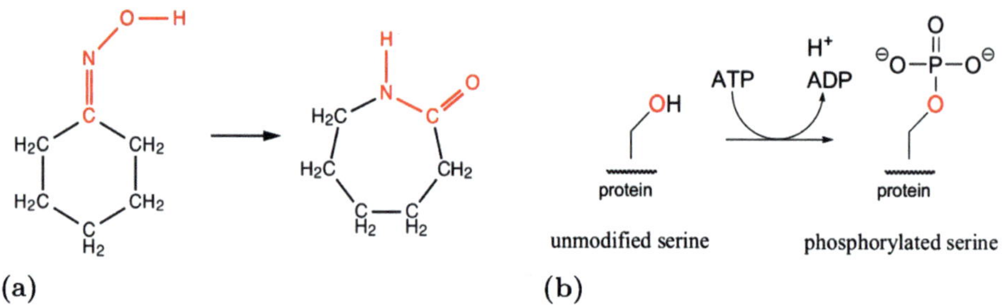

Figure 2.1. *An example chemical reactions.*

Source: Trojak, M., Šafránek, D., Mertová, L., & Brim, L., (2020). Executable Biochemical Space for Specification and Analysis of Biochemical Systems. – Scientific Figure on ResearchGate. Available from: https://www.researchgate.net/figure/Examples-of-chemical-reactions-A-An-example-of-elemental-chemical-reaction-The_fig1_344231693.

After learning this shocking information, more than 86% of his classmates signed a petition to make it illegal. They had no idea, however, that dihydrogen monoxide is simply water. In fact, drowning is fatal; its steam causes severe burns; and tumor cells contain more water than healthy cells. These are all true facts that were not made up. Because people did not consider writing the chemical formula for dihydrogen monoxide from its name, H2O, the prank was successful. If they had taken the time to write down the formula, they would not have fallen into Nathan's trap. Nathan's winning entry in the competition was titled "How gullible are we?"

If we don't want to fall victim to such practical jokes, we must learn how to write chemical formulae and reactions. Let us go deep together!

2.2. CHEMICAL FORMULA DEFINITION

A chemical formula, which is a collection of chemical symbols, represents the number or ratio of atoms of each type of element in a molecule or compound (Fleming, 2011). From the preceding account, how can you deduce the chemical name of water, dihydrogen monoxide?

'Di' means two, and 'mono' means one. As a result, the elements monoxide and dihydrogen both indicate the presence of one oxygen atom and two hydrogen atoms. H_2O is the chemical formula for water.

2.2.1. Types of Chemical Formulae

There are 3 types of chemical formulas involves in chemical reactions: Molecular formula, empirical formula and displayed formula (Davis, 2012).

2.2.1.1. Molecular Formula

The molecular formula indicates the number of each type of atom in a compound or molecule. For example, the chemical formula for glucose is $C_6H_{12}O_6$. A glucose molecule has six carbon atoms (C6), 12 hydrogen atoms (H_{12}), and six oxygen atoms (O_6), according to this formula (Fleming, 2011) (Figure 2.2).

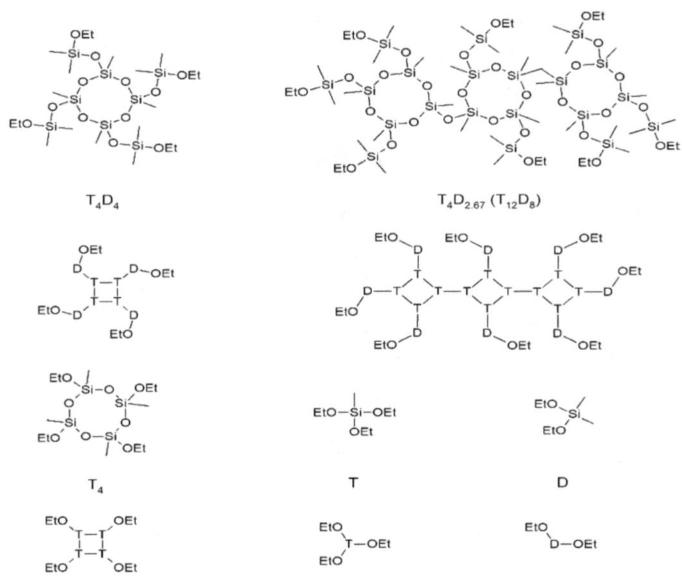

Figure 2.2. *Example of molecular formulas.*

Source: Owińska, et al., (2021). Macromonomers as a Novel Way to Investigate and Tailor Silicon-Oxycarbide-Based Materials Obtained from Polymeric Preceramic Precursors. – Scientific Figure on ResearchGate. Available from:

https://www.researchgate.net/figure/Molecular-formulas-denoted-at-the-top-and-structural-unit-compositions-depicted-below_fig1_352956528.

2.2.1.2. Empirical Formulae

The empirical formula of a compound displays the simplest whole-number ratio of each element's atoms that are present in the compound or molecule.

For example, the empirical formula for glucose is CH_2O. Glucose has exactly twice as many hydrogen atoms as carbon and oxygen atoms, according to this equation. In other words, the ratio of C, H, and O atoms in glucose is 1:2:1. We always use empirical formulae to represent ionic compounds. Because ionic compounds can have an infinite number of constituents, they lack a fixed molecular formula.

2.2.1.3. Displayed Formula

Every atom and bond in a molecule can be seen in the compound's formula. Each bond corresponds to a pair of electrons in the displayed formula. $C_6H_{12}O_6$ is the molecular formula for glucose and fructose. They clearly share the same empirical formula, which is CH_2O. So, how do we tell the difference between the two?

2.2.2. Chemical Formulae Writing Rules

The fundamental concept of chemistry is learning how to express chemical formulae in writing. Writing formulas can help you by many aspects of chemistry, such as creating balanced chemical equations and calculating the masses of molecules and compounds. In other words, everything you learn after this is dependent on your ability to write chemical formulas correctly using the elemental symbols you've learned (Davis, 2012).

When writing chemical formulas, two important factors must be considered:

- Because ionic compounds are not actual molecules, they lack a molecular formula. In ionic compounds, which are giant lattices (for example, NaCl), ionic formula units are represented by empirical formulae.
- Because covalent compounds are actual molecules, they are represented by chemical formulas.

There are a few rules to follow when writing chemical formulas (Fleming, 2011):

- Each element's chemical symbol is displayed.
- The charges of the ions should be predicted based on their position in the periodic table.
- The sum of the charges of the ions in an ionic species must be zero.
- In covalent species, the partial charges of the elements must balance.
- The number of atoms in each element or ion is indicated in the formula to the right of its symbol in subscript.
- When a compound contains multiple polyatomic ions, the subscript for each ion is written to the right of the closing bracket (for example, $Cu(SO_4)_2$).
- Brackets are not required for polyatomic ions with a ratio of one (for example, NaOH and NH_4NO_3).
- The name of the cation is always written first in ionic species. This means that if a compound contains both a metal and a nonmetal, the symbol for the metal must come first, followed by the symbol for the nonmetal.
- Transition metals with two or more distinct charges can combine with the same nonmetal to form a wide range of compounds. Copper and Chlorine, for example, can react to form the compounds cupric chloride ($CuCl_2$) and cuprous chloride (CuCl). The Roman numeral in parentheses after the corresponding metal ion, in this case, copper, denotes the charge of copper in cuprous chloride, which is +1. Copper(II) chloride is written in the same way that cupric chloride is.

The most prevalent polyatomic ions that one will encounter are mentioned below:

Name of the Polyatomic Ion	**Example**
Nitrate	Sodium nitrate-
Nitrite	Sodium nitrite-
Carbonate	Sodium carbonate
Sulfate	Sodium sulfate
Sulfite	Sodium sulfite
Phosphate	Sodium phosphate-

Phosphite	Trisodium phosphite
Phosphide	Sodium phosphide
Hydroxide	Sodium hydroxide
Ammonium	Ammonium chloride
Bicarbonate/Hydrogen carbonate	Sodium bicarbonate
Bisulfate/ Hydrogen sulfate	Sodium hydrogen sulfate
Bisulfite/ Hydrogen sulfite	Sodium hydrogen sulfite
Permanganate	Potassium permanganate

2.2.2.1. The Procedure for Writing Chemical Formulas

Consider the example of copper(II) chloride to refer how to write a chemical formula. Refer to the standards for writing chemical equations to remember that copper and chlorine can produce two different compounds.

List the elements or ions present in the compound before attempting to build the formula. As an example, consider the simple compound copper(II) chloride. This mixture contains both copper and chlorine, as we can see.

Next, write down the chemical symbols for the elements and ions you identified in the previous step. If the compound is made up of a metal and a nonmetal, start with the metal's symbol, such as Cu and Cl.

Directly beneath each of the chemical symbols for the atoms and ions in the molecule, write the charges of the individual ions. In this case, –1 for chlorine and +2 for copper (Fleming, 2011).

Now, cross the valencies of each element with the matching element or ion. Finally, the charge of one ion determines the ratio or subscript of the other element or ion. Remember to subscript the ratio.

When atoms combine to form molecules, gain or lose electrons to form ions, or combine with other atoms to form ions, chemical formulas are created that accurately describe the new species. To extend this symbolism to represent both the identities and proportional amounts of substances which undergo a chemical (or physical) change, a chemical equation must be written and balanced (Davis, 2012).

As an example, consider the reaction of one molecule of methane (CH_4) and two molecules of oxygen gas (O_2). When these two substances react, one molecule of CO_2 and two molecules of water (H_2O) are produced (Figure 2.3):

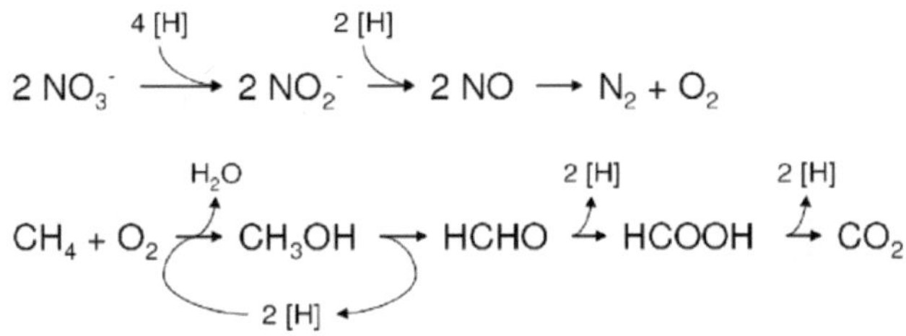

Figure 2.3. *An illustration of methane reaction with oxygen followed by denitrification.*

Source: Deutzmann, J., (2011). Aerobic and Anaerobic Oxidation of Methane in Sediments of Lake Constance. – Scientific Figure on ResearchGate. Available from: https://www.researchgate.net/figure/Reaction-scheme-of-methane-oxidation-coupled-to-denitrification-with-oxygen-as_fig15_283271184.

The following example shows the fundamental elements of any chemical equation:

- The substances that undergo a reaction are referred to as reactants, and their formulas are listed on the left side of the equation.
- On the right side of the equation, the formulas for the products, which are the substances produced by the reaction, are listed.
- Individual reactant and product formulas are denoted by plus signs (+), and the equation's reactant and product (left and right) sides are denoted by an arrow ().
- The relative numbers of reactant and product species are represented by coefficients (numbers to the left of each formula). A coefficient of one is usually omitted.
- In a chemical equation, it is customary to use as few whole-number coefficients as possible, as shown in this example. However, keep in mind that these coefficients are actually ratios because they represent the relative amounts of reactants and products. Methane and Oxygen react in a 1:2:1:2 ratio to produce CO_2 and water. This ratio is satisfied if the numbers of these molecules are 1-2-1-2, 2-4-2-4, 3-6-3-6, and so on. Because these coefficients can be interpreted in relation to any amount (number) unit, this equation can be correctly read in a variety of ways.

- The reaction between one methane molecule and two oxygen molecules produces one CO_2 molecule and two water molecules (Davis, 2012).
- 12 methane molecules combine with 22 oxygen molecules to form 12 CO_2 molecules and 22 water molecules.
- The reaction between one mole of methane (6.02 x 1023 molecule) and two moles of oxygen (1.204 x 1024 molecules) produces one mole of CO_2 and two moles of water (Fleming, 2011).

2.2.3. More Information on Chemical Equations

To indicate the physical states of the reactants and products, chemical equations frequently include an abbreviation in parentheses after the formulas. Some common abbreviations are S for solids, l for liquids, g for gases, and aq. for substances dissolved in water (aqueous solutions, as introduced in the previous chapter). These notations are illustrated in the following example equation:

$$2Na(s)+2H_2O(l) \longrightarrow 2NaOH(aq) + H_2(g)$$

This equation represents the reaction that occurs when sodium metal is mixed with water. When solid sodium reacts with liquid water, the ionic compound sodium hydroxide is formed, which is solid in pure form but easily dissolved in water.

Writing a word or symbol above or below the arrow of an equation can sometimes be used to designate special conditions required for a reaction. For example, the uppercase Greek letter delta (Δ) over the arrow may denote a heating reaction.

$$CaCO_3(s)\Delta- \longrightarrow CaO(s) + CO_2(g)$$

If you frequently work with written equations, learning the symbols for these situations can be beneficial.

2.2.4. Ionic Substances in a Solution

When NaCl dissolves in water, the ions separate and go their separate ways in solution; they are now identified by their respective charges, and the (aq.) phase label emphasizes that they have dissolved. This is referred to as dissociation; the ions are said to dissociate (Figure 2.4).

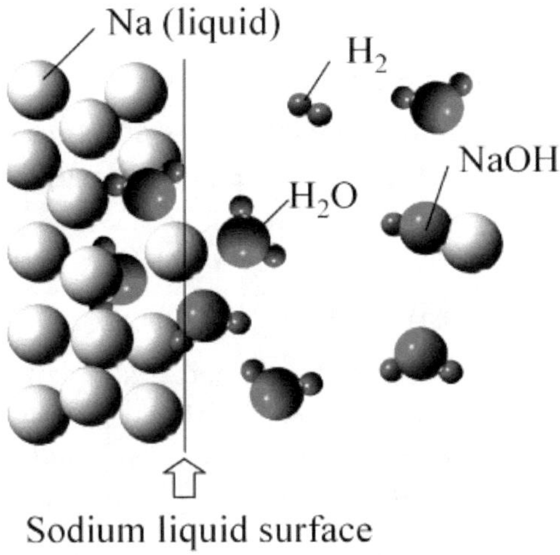

Figure 2.4. *Sodium and water reaction.*

Source: Takata, T., Yamaguchi, A., Fukuzawa, K., & Matsubara, K., (2005). Numerical methodology of sodium-water reaction with multiphase flow analysis. Nuclear Science and Engineering. – Scientific Figure on ResearchGate. Available from: https://www.researchgate.net/figure/Schematic-of-the-surface-reaction_fig1_236381415.

When an ionic compound dissociates in water, it is surrounded by water molecules, which separate it from the rest of the compound. Every ion in a solution follows its own path (Davis, 2012).

Remember that when the ions separate, every ion separates. As a result, when $CaCl_2$ dissolves, the two Cl ions and the one Ca^{2+} ion separate:

$CaCl_2(s) \longrightarrow Ca^{2+}(aq) + 2Cl^-(aq)$

or

$CaCl_2(s) \longrightarrow\longrightarrow Ca^{2+}(aq) + Cl^-(aq) + Cl^-(aq)$

After that, the two chloride ions split on their own. They do not combine to generate Cl_2 or $2Cl^-$, which would be elemental chlorine; instead, they form chloride ions. They are dissociated in order to create ions. Polyatomic ions retain their general identity even when dissolved (Fleming, 2011).

Illustration 2 write down the chemical formula for each ionic compound's dissociation.

i. KBr;

ii. Na_2SO_4.

Solution:

i. $KBr(s) \longrightarrow K^+(aq) + Br^-(aq)$

ii. In addition to the two sodium ions splitting off, the sulfate ion remains in its original state. The breakdown formula is as follows:

$Na_2SO_4(s) \longrightarrow 2Na^+(aq) + SO_4^{2-}(aq)$

2.2.5. Ionic Reactions Equations

Given the abundance of water on the planet, it stands to reason that many chemical reactions take place in water solutions. Depending on how many ions are involved in the reaction and what their intended use is, the chemical equations can be written with varying levels of detail. As an example, consider an ionic compound reaction that occurs in an aqueous solution. When aqueous solutions of $CaCl_2$ and $AgNO_3$ are combined, a reaction occurs that produces aqueous $Ca(NO_3)_2$ and solid AgCl:

$CaCl_2(aq)+2AgNO_3(aq) \longrightarrow Ca(NO_3)_2(aq)+2AgCl(s)$

This balanced equation, which was derived in the usual way, is known as a molecular equation because it does not explicitly represent the ionic species present in the solution. Ions in ionic compounds that dissolve in water can dissociate into individual ions, which are then uniformly distributed throughout the solution. As a result, in this case, dissociated ions are a more accurate representation of ionic compounds dissolved in water:

$CaCl_2(aq)Ca^{2+}(aq)+2Cl^-(aq)$

$2AgNO_3(aq)2Ag+(aq)+2NO^{3-(}aq)$

$Ca(NO_3)_2(aq)Ca_2^+(aq)+2NO^{3-}(aq)$

AgCl, which has physical state notations, does not dissolve in water as much as the other three ionic compounds (Davis, 2012).

When all dissolved ions are explicitly represented, the ionic equation is complete. In this case, the dissolved ionic compound formulas are replaced by dissociated ion formulas:

$Ca^{2+}(aq)+2Cl^-(aq)+2Ag^+(aq)+2NO^{3-}(aq) \longrightarrow Ca^{2+}(aq)+2NO^{3-}(aq)+2AgCl(s)$

Examining this equation reveals that two chemical species, $Ca^{2+}(aq)$ and $NO^{3-}(aq)$, are present in the same form on either side of the arrow. $NO^{3-}(aq)$. Because the process does not change these spectator ions, whose presence is required to maintain charge neutrality, they can be removed from the equation to produce a more concise illustration known as a net ionic equation:

$Ca^{2+}(aq)+2Cl^{-}(aq)+2Ag^{+}(aq)+2NO^{-3}(aq) \longrightarrow Ca^{2+}(aq)+2NO^{3-}(aq)+2AgCl(s)$

$2Cl^{-}(aq)+2Ag^{+}(aq) \longrightarrow 2AgCl(s)$

This equation is then written with the smallest integers as coefficients, as is customary:

$Cl^{-}(aq)+Ag^{+}(aq) \longrightarrow AgCl(s)$

This net ionic equation states that silver(I) ions from any source can be used to produce solid silver chloride from dissolved chloride. These molecular and complete ionic equations, which also provide additional information, detail the ionic compounds used as sources of Cl^{-} and Ag^{+}.

2.2.6. Soluble and Insoluble Ionic Compounds

The concept of solubility versus insolubility in ionic compounds is a matter of degree. The solubility of ionic compounds ranges from highly soluble to moderately soluble to practically insoluble. For the majority of ionic compounds, there is also a maximum amount of compound that can dissolve in a sample of water. For example, at room temperature, 36.0 g of sodium chloride can be dissolved in 100 g of water, but only 0.00019 g of silver chloride. In comparison to NaCl, AgCl is thought to be insoluble in water.

One application where solubility is critical is the tank-type water heater found in many American homes. Domestic water contains trace amounts of dissolved ionic compounds such as calcium carbonate ($CaCO_3$). $CaCO_3$ has an unusual property in that it is less soluble in hot water than in cold water. Because the water heater works by heating water, $CaCO_3$ can precipitate if there is enough of it in the water. Limescale is a precipitate that may contain phosphate, magnesium, and hydrogen carbonate compounds. The problem is that excessive limescale can cause a water heater to malfunction by requiring more energy to heat water to a specific temperature or even obstructing the water heater's entry or exit pipes (Fleming, 2011).

Another location where solubility versus insolubility is an issue is the Grand Canyon. Rocks are typically thought to be insoluble. However, it is only very slightly soluble. This means that the Colorado River slowly dissolve rocks from the surface over roughly 2 billion years, eventually creating a stunning array of gorges and canyons. And it's all because of solubility! Chemical equations represent chemical changes symbolically. An arrow separates the formulas for the substances change c (reactants) and those produced by the change (products), and integer coefficients preceding the formulas indicate the relative number of the two groups of substances. Because of their coefficients, balanced equations produce an equal number of atoms for each element in the reactants and products. Complete ionic equations and more succinctly, net ionic equations can more accurately depict chemical reactions involving ionic reactants or products in aqueous solutions (Davis, 2012).

2.3. READING CHEMICAL EQUATIONS

It is recommended that you read the topic Reading chemical formulas before beginning this topic. Chemists use a special shorthand to write chemical names and their behavior in chemical reactions.

Water, for example, has the chemical formula H_2O, and when airborne oxygen and aluminum combine to cause corrosion, chemists use the following formula:

$$4Al + 3O_2 \longrightarrow 2Al_2O_3$$

2.3.1. Equations and Chemical Reactions

Chemists have formulas or shorthand names for every known atom, ion, molecule, and ionic salt. These formulas list the atom types and numbers of each atom present in a substance.

During chemical reactions, substances are destroyed and new ones are created. Once the substances involved in a reaction have been identified, a chemist can use the valency and covalency rules to determine the formula of the substances. These regulations establish a constant formula for a substance. Water has the chemical formula H_2O at all times.

"The reactants" are the substances that start a chemical reaction, and "the products" are the new substances that emerge as a result of the reaction. When an arrow in the middle of a chemical equation with the reactants on the left and the products on the right, a reaction has occurred.

If there are multiple reactants or products produced, plus signs (+) are used between the formulas of the substances involved. Product 1 multiplied by Product 2 equals Add Reactant 1 and Reactant 2.

In the first example of aluminum corrosion, aluminum, and oxygen are the reactants and aluminum oxide is the sole product.

$$4\ Al + 3\ O_2 \longrightarrow Al_2O_3$$

What Do the Numbers Mean?

Keep in mind that subscript numbers, such as the 2 in O_2, are part of the formulas and indicate how many atoms of each type are present in the formulas of the substance. The larger numbers in front of the chemical formulas represent the amount of each atom or compound that underwent reaction or was formed as a result of the reaction and is a part of the chemical equation (Fleming, 2011).

At its most basic level, the equation $4Al + 3O_2$ gives Al_2O_3 can be translated as "Four aluminum atoms react with three oxygen molecules to form two units of aluminum oxide."

- Ø **Note:** Just like salt or sand, aluminum oxide exists as crystals in a lattice. The components of aluminum oxide cannot technically be referred to as molecules.

Atoms are consistently displayed as simple elementary formulas without any subscripts. Examples include H, He, Al, C, and Ca. O_2, H_2O, and $C_6H_{12}O_6$ are a few examples of molecules that are made up of two or more atoms.

The distinction between covalent molecules and ionic salts is not covered here, as it has been done with all topics at a fundamental level (Davis, 2012).

Following are two additional equations and their results:

$$2H_2 + O_2 \longrightarrow 2H_2O$$

"Two water molecules are created when two hydrogen atoms combine with an oxygen molecule."

The equation states that when lime, calcium carbonate, $CaCO_3$, is heated, it transforms into CO_2, and calcium oxide, heat.

"A unit of calcium carbonate decomposes on heating to form a unit of calcium oxide and a molecule of carbon dioxide," according to the formula $CaCO_3$ gives $CaO + CO_2$.

- Ø ***Remark:*** Crystals of calcium carbonate and calcium oxide can be found in a lattice. It is technically false to refer to calcium carbonate and calcium oxide as molecules.

The only other characteristic of an equation that is frequently used is the nature of the constituents, whether they are in a solid, liquid, gas, or solution of water.

A subscript (s) is added to the substance's chemical formula, CaO(s), to indicate that it is solid.

A subscript (l) is placed after the substance's chemical formula, $H_2O(l)$, to indicate liquid.

Following the substance's formula, $CO_2(g)$, a subscript (g) designates the substance as a gas.

NaCl(aq)'s chemical formula is followed by the subscript (aq), which designates water solutions as "aqueous solutions."

As a result, when lime is heated:heat

$CO_2(g) + CaCO_3(s) + CaO(s)$

A unit of solid calcium carbonate decomposes into a unit of solid calcium oxide and a gaseous CO_2 molecule on heating"

When water forms, it becomes:

$O_2(g) + 2H(g) \longrightarrow 2H_2O(l)$

Two molecules of liquid water are created when two molecules of gaseous hydrogen and one molecule of gaseous oxygen interact (Fleming, 2011).

2.4. PRINCIPLES OF CHEMICAL REACTION

2.4.1. Kinetics and Reaction

Mechanism Without a doubt, being able to directly observe the activities of individual atoms and molecules to see what they can and cannot do would be extremely beneficial in understanding what reactions are possible. Even though such observations will never be possible, experimental measurements of the reaction rates of the substances in bulk allow us to deduce what they must be. Kinetics research can then provide useful information about the activation energies required for reactions as well as the mechanisms by which they are most likely to occur. However, such knowledge is primarily descriptive, leaving unanswered the fundamental question of why the atom arrangement in the products should be preferred over that in the reactants. Understanding the activation energy required, for example, can help explain why a given reaction may proceed quickly or very slowly under certain conditions (Fleming, 2011).

Unexpected Reaction Characteristics As a result, rather than focusing on the rates or mechanisms of the various transition processes, the nature of the initial and final states is our primary concern in this case. Because only the characteristics of these states can determine which is more likely to be the result of one reacting spontaneously to the other. The term "spontaneous" must be understood in context. It is frequently used to describe reactions that begin immediately when reactants come into contact with one another. A substance is called "spontaneously inflammable" if it spontaneously ignites when exposed to oxygen. Although such instances are unmistakably spontaneous, they represent only a small proportion of what is more commonly referred to as "spontaneous" reactions. In its most literal sense, "spontaneous" means "having the potential to proceed without the assistance of an external agency." When a reaction is referred to as spontaneous, no inferences are made about its mechanism or rate, and no indication is given as to how or whether it will proceed under the current conditions. Nonetheless, the direction of spontaneity remains the most important characteristic of a chemical reaction (Davis, 2012).

As previously stated, the relative propensity of separated atoms to come together in a particular way is determined by two factors. The first is the quality of the bonds formed. The other factor is the likelihood or level of randomness in the outcome. Stronger bonds and a disorganized array tend to cause a spontaneous reaction. The thermodynamic function known as the "Gibbs energy," abbreviated G, takes these two influences on the spontaneity of chemical change into account. At constant pressure, the "net energy" is the difference between the total energy content, or enthalpy, and the energy that is not available for useful work, or entropy S, and the absolute temperature, T. The formula G=H-TS represents this function.

Entropy is calculated using the energy change per degree, with "entropy units" equaling the number of calories per degree per mole. Multiplying by the absolute temperature yields the energy associated with the probability at that temperature, or the energy unavailable for useful work. Changes in free energy, heat content, and entropy are the quantities of interest in a chemical reaction: TAS – AH = AG As a result, the first and foremost fundamental principle of chemical reaction is:

Because chemical reactions tend to proceed spontaneously in the direction of decreased free energy content, free energy is released when the free energy change, AG, for the reaction is negative.

According to the expression, only one of the following sets of circumstances will result in a chemical reaction occurring spontaneously at constant pressure: If the total disorder (entropy) of the products is higher and their total bonding forces are greater than those of the reactants; if the total bonding forces in the products are greater than the total bonding forces in the reactants, but not by enough to raise TAS above AH; or If the overall bonding forces between reactants and products are weaker, but the increase in entropy (or disorder) is greater than the heat absorbed.

Where these two influences are at odds, the ha[1] result will be determined by the relative contributions of bond strength and entropy to the reaction. We can now investigate more closely at the variables influencing the sizes of the relative contributions.

2.4.2. Principles of Entropy

The gaseous state is more likely than the liquid state, which is more likely than the solid state. In other words, because atoms or molecules in a gas are independent of one another than they are in a liquid or solid, they can become more randomly distributed, resulting in higher entropy. Entropy increases during melting and even more in vaporization. Because the motion of the molecules in the liquid state is still very constrained, the entropy of melting is relatively low (Fleming, 2011).

Because of the significant difference in orderliness between liquid and gas, the entropy of vaporization is relatively high. For "normal" liquids, or those with van der Waals forces as the primary intermolecular attraction, the entropy of vaporization is approximately 22 entropy units. Because the vaporization process necessitates a greater drive toward disorder, stronger polar interactions or protonic bridging in the liquid will increase entropy. The melting entropy, which for a large number of chemical elements is equal to 2.3 entropy units, is also essentially constant.

According to this principle, the presence of gas among the products but not among the reactants indicates that the entropy of the products is most likely greater than that of the reactants. This tendency occurs because the entropy differences between different solids and liquids are typically less than those between the gaseous and solid or liquid states. As a result, changes in entropy frequently favor the breakdown of solids or liquids into gaseous products (Davis, 2012).

Monoatomic gases have higher entropy than polyatomic gases because they are more likely to exist. The monatomic gas is in a state of greater disorder.

Polyatomic molecules, with their fixed and definite structure, help to create more order, which reduces entropy. In a gaseous reaction, the change in entropy favors any metathesis (double decomposition) that increases the total number of moles of gas or the dissociation of larger molecules into smaller molecules.

An amorphous solid is more likely than a crystalline solid, and a simple crystalline solid is more likely than a more complex crystalline solid. As a result, the entropy change favors the formation of simpler substances from more complex ones.

A molecular addition compound, also known as a coordination complex, has a lower probability of occurrence than its constituent parts. As a result, the formation of such compounds is accompanied by a decrease in entropy, which raises the bond strength requirement required to maintain aggregate stability.

Higher atomic weight elements tend to form molecules or compounds with higher entropy levels. The explanation is thought to be the effect of mass on intra-crystalline or intra-molecular vibrations. With increasing mass, the vibration frequency decreases, making the state more likely and increasing entropy.

Entropy effects at ordinary temperatures are typically small enough to have little impact on the direction of the reaction unless the difference in total bonding energy between reactants and products is relatively small. This is effectively demonstrated by comparing the AH and AG values for the formation of several chemical compounds in their standard. This fact, which occurred before the concept of entropy was fully understood, led Berthelot and Thomsen to believe that the only thing dictating the direction of spontaneous reaction was the heat of the reaction. Heats of reaction are typically greater than one kilocalorie per mole, whereas entropy units are calories per degree per mole, and entropy changes associated with reactions are frequently only a few entropy units. Even though the entropy is multiplied by approximately 300°K (at normal temperatures), the result does not even equal one kilocalorie unless the entropy is greater than 3. If the AH value for the reaction is relatively high, say – 10 kcal per mole or higher, one can be reasonably confident in predicting a spontaneous reaction in that direction (Fleming, 2011).

All chemical reactions that increase entropy occur spontaneously at sufficiently high temperatures. Even though the heat of the reaction changes relatively slowly with temperature, the entropy effect is the result of entropy and temperature. If the entropy change is positive and multiplied by a large enough temperature, the product TAS will be subtracted from the heat of the reaction to produce a negative free energy change. This is an important principle explaining why species unknown at lower temperatures can exist at high temperatures due to their proclivity to form more complex aggregates stable at lower temperatures. Compound decompositions, as previously stated, tend to occur spontaneously at high enough temperatures because they increase entropy. While enthalpy changes typically control the course of reactions at normal temperatures, as the temperature rises, entropy's relative importance rises until it overtakes all other factors.

2.4.3. Principles of Bond Strength

Covalence on a regular basis It is critical to distinguish between "solubility" and "reactivity." Unfortunately, people frequently confuse the two. Even if its bonds are extremely strong, a substance that is relatively reactive is far too frequently labeled as "unstable."

Given this ambiguity, it appears far preferable to limit the definition of "stable" to the state of being held together by strong bonds and to suggest that a high temperature is required to disrupt the substance, regardless of its potential chemical reactivity. Then, regardless of how reactive or inert a substance is, it is said to be "unstable" if it has weak bonds that are thermally broken at relatively low temperatures.

It is prudent to consider the variables that may influence bond strength. Unfortunately, due to the topic's complexity, it can only be studied qualitatively or semi-quantitatively at the moment. However, some useful generalizations can be made. In fact, it was recently discovered that by applying the following principle, it is possible to calculate the total bonding energy quantitatively for at least some types of compounds, particularly binary halides:

Total bonding energy (per equivalent) varies inversely with bond length and directly with halogen partial charge in binary halides and possibly other classes of compounds. The fact that all of the major group elements' binary chlorides fall on the same straight line connecting Cl with NaCl, which is formed by plotting partial charge on chlorine against the function RH/n (where n is the bond order, H is the heat of atomization per equivalent, and R is the bond length), shows that this principle holds true regardless of the type of bonding. A bonding theory to explain these remarkable relationships is still a long way off, but combining "covalent" and "ionic" bonding, eventually promises to enable the more thorough understanding required for more accurately predicting the direction of spontaneous change. Returning to more traditional notions of bonding, we may notice some guidelines that are useful in the interim (Davis, 2012).

2.4.4. Normal Covalence

Covalent bonds formed between small atoms are typically stronger than those formed between larger atoms. Given the importance of coulombic forces in covalence, closer interactions between opposite charges are likely to result in stronger bonds between smaller atoms because these forces vary inversely with the square of the distance between the charges. Larger atoms have more non-bonding electrons, which have the potential to weaken bonds through inter-cloud repulsions in addition to longer bonds. This effect of appeal outweighs the effect of attraction caused by the increased nuclear charge (Fleming, 2011).

The greater the difference between the principal quantum numbers of the two elements' valence shells, the weaker the covalent bonds between atoms. When the atoms are very disproportionally sized, the overlapping or coalescence of orbitals is less effective. Binary hydrogen compounds with relatively low polarity covalent bonds are a good example of this.

The stronger the covalent bonds, the higher the bond polarity. Pauling proposed one explanation for this tendency, believing that the extra energy caused by bond polarity resulted from the combination of covalent bond energy and ionic attraction between opposite charges. Using the extra energy, he created his famous electronegativity scale. Shorter polar bonds appear to allow for closer interaction between bonding electrons and nuclei. The unequal electron distribution causes a radial contraction of the partially positive atom that is greater than the radial expansion caused by the other atom acquiring a partial negative charge. The heats of atomization (per equivalent) of binary compounds with different oxidation states provide a rough quantitative indication of the impact of polarity differences in bonds between the same atoms. The heat of atomization always decreases as the positive oxidation state of the central atom increases. This is because the more atoms competing for a given atom's electrons, the less successful each will be at acquiring a negative charge, and thus the less polar its bond.

Bonds between hydrogen and highly electronegative elements like fluorine, oxygen, and chlorine are especially strong because the relatively positive hydrogen atom lacks any underlying electron shells to limit the closeness of interaction with the other atom. The significance of this additional strength in influencing the course of the reaction will be discussed later (Fleming, 2011) (Figure 2.5).

Covalent bonds between atoms with a small positive charge are typically less reliable. On the contrary to what might be expected from the fact that positive charge corresponds to contracted electronic spheres, such bonds tend to be as long as expected from their nonpolar covalent radii. This generally translates to bond weakening and may be caused by a reduction in the number of outer electrons available for sharing.

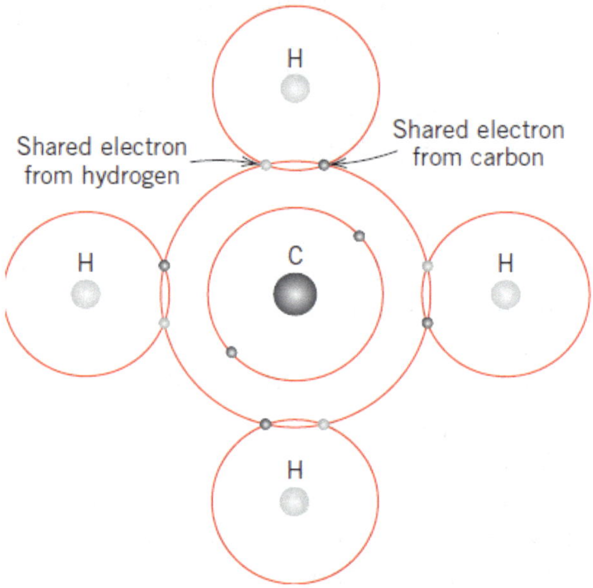

Figure 2.5. *An illustration of covalent bond.*

Source: Abdullah, Y., & Abdal-Rhman, (2019). Chemical Bonding an Overview. – Scientific Figure on ResearchGate. Available from: https://www.researchgate.net/figure/Schematic-representation-of-covalent-bonding-in-a-molecule-of-methane_fig2_331546390.

Covalent bonds with partial multiplicity as a result of resonance are more stable than single bonds or the equivalent partial multiplicity in separate single and integral multiple bonds. The validity of this principle appears to be supported by resonance, which occurs when bonds are equalized toward partial multiplicity rather than distinct single and double bonds. Perhaps the answer can be found in the apparent increase in interelectronic repulsions caused by an atom gaining electrons. This rise appears to indicate that shared electrons, which are initially acquired in part, contribute the most to stability. As a result, holding the bonding electrons partially rather than completely results in greater stability between two (or more) atoms. This explains how covalent bonds form and why they require energy to form (Davis, 2012).

Similarly, a double bond that can only be in one of two places in a molecule is more likely to be distributed partially in both places rather than entirely in one. This type of situation, known as a resonance, occurs wherever it is possible because it creates more stability. We can actually anticipate further interaction in which these orbitals and electron pairs are used at least partially to strengthen the bonding through multiplicity wherever formation of single covalent bonds would leave at least one otherwise vacant orbital on one atom and one or more pairs of otherwise unshared outer electrons on adjacent atoms (Fleming, 2011).

Fluorine and oxygen compounds with other elements whose atoms have open outer vacant orbitals are the most notable examples of bonds where partial multiplicity is to be expected. Fluorine is thought to have the highest electronegativity of any element and only one open valence orbital, so it can only form one covalent bond. However, according to a number of facts, such an bond could be partially multiple under the right conditions. The fluoride ion is known to function as the donor in many complexes; partially negative fluorine may also function in this capacity, albeit to a lesser extent. Nonbonding outer electron pairs on the two atoms interfered with one another, according to one explanation for the weakening of the bond in Fl, implying that such pairs could interact favorably with another atom aif it had open orbitals.

Another sign of partial multiplicity is the frequent formation of disproportionately strong and unexpectedly short bonds between other elements and fluorine. The fluorides of other highly electronegative elements show bond lengths that are consistent with what would be expected if the covalent radii were added; however, for less electronegative elements, the calculated radius sum frequently needs to be divided by 0.94 to produce the observed bond length, indicating a bond order of approximately 1.5. The analogy to oxygen, which forms two bonds and reveals information about its bonding via the bond angle, lends support to the partial multiplicity in fluorine bonding. Although two single bonds to oxygen should form an angle of no more than 109' (ClO, 110.8'; FlO, 103.2"; HIO, 104.5'), there are numerous instances where the angle is much larger, even close to 180.' For example, the Si_4Si angle in many silicates and silicones is between 140 and 150,' the P-0-P angle in PnOs, As, and Sb-O_SH is between 12 and S0, and the B-0-B angle in a variety of borates is between 120 and 140.' All of these unusually large angles point to partial utilization of the oxygen atoms two pairs of outer electrons, possibly involving the outer d orbitals of silicon, phosphorus, or the fourth spa orbital of boron. There appears to be no reason why fluorine cannot participate in such partial multiplicity in the same way that oxygen can.

Heats of atomization (per equivalent) of BFI (154.5), B_20 (126.9), AFI (148.5), AlaOs (121.9), SiFl (137.8), SiO_2 (107.2), AsF_3 (114.9), and similar compounds are significantly higher than polarity would predict, most likely due to such partial multiplicity. When the average bond energies are arranged in descending order, the gaseous molecules listed in NBS Circular 500 (8) have the following leaders: B-F, Si-F, H-F, Li-C1, Na-F, As-F, and H-4. As will be seen, these facts are especially important when attempting to predict how such compounds will react (Davis, 2012).

2.4.5. Coordination

This topic could easily lead to far more in-depth and drawn-out discussions than are necessary for our main goal, which is to establish useful guidelines for understanding chemical reactivity. Such discussions would concentrate on the chemistry of transition metal complex ions and expose ligand field theory in greater depth and at a higher level of specialization than this paper intended. Despite being a fascinating and active area of research, it is not included on this list. Let us instead concentrate on some of the more fundamental aspects of coordination bond formation, which affect both complex ions and molecules with molecular additions (Fleming, 2011).

The stronger the coordination bonds are, the more easily the donor atom can supply electrons. The following factors may influence an atom's ability to provide a pair of electrons to an acceptor:

Its Partial Charge, as shown in (a). An atom that has some of its electrons withdrawn will not readily donate an electron pair. For example, neither nitrogen trifluoride nor (CFN, in which the nitrogen atom has a partial positive charge) are good donors. Even if the donor atom is of a highly electronegative element and carries only a minor partial negative charge in the compound, it is still ineffective even if other electrons are not removed from it. The oxygen in nitrate or sulfate ions, for example, is not very negatively charged and does not readily donate electrons.

On the other hand, even if the donor atom had a high initial electronegativity, if it is highly negative, it can still work well as a donor. When oxygen has a relatively high negative charge, it can be used as a donor; for example, hydroxide ions are more effective donors because they have an oxygen charge of – 0.67 rather than water's – 0.25. When two or more elements combine to form a negative ion, the unshared electrons in the outer shell of the original element are expected to donate more readily. In the cyanide ion, for example, both carbon and nitrogen receive unshared electron pairs, but carbon is likely the more willing donor. Nitrogen is more likely to donate in anions that have both nitrogen and oxygen as potential donors. In compounds with even a minor partial positive charge, such as $P(CH_3)_3$, the original element, which is only intermediate in electronegativity, can act as a donor, especially if the attached atoms or groups are good electron suppliers. The discovery that a donor's coordination number in a crystal lattice increases as donor negative charge increases:

1. Donor Involvement in Bonding. This donor ability is closely related to multiplicity. When an electron pair is involved in multiple bonding via resonance, it becomes less available for coordinate bond formation. Because silicon is less electronegative than carbon, the partial charge on nitrogen in trisilylamine [$Si(CH_3)_3N$] should be more negative than in trimethylamine. Trisilylamine should be a better donor as a result, but it is far from efficient. The "unshared" pair of electrons on nitrogen can become involved in the outer orbitals of silicon to the point where they are effectively used in the nitrogen-silicon "pi" bonding and thus unavailable for donor action toward an external acceptor.
2. Reservoir capacity in relation to the donor. The larger the electron reservoir made available to donor atoms by other atoms covalently bonded to them, the better their ability to supply electrons appears to be. $P(CH_3)_3$ for example, is a better donor than PH_3 because the methyl groups can supply the phosphorus atom with electrons more efficiently than the hydrogen atoms can, allowing the phosphorus atom to share its extra pair of electrons with an external acceptor (Davis, 2012).
3. The donor's polarizability. When compared to the fluoride ion, the iodide ion may be more effective in some combinations if the donor is relatively easy to polarize (Fleming, 2011).
4. The concentration of the donor electron pair. Donor atoms may be better able to function when electron pairs are concentrated within a relatively narrow region rather than spread out, especially toward small atom acceptors. Alcohols and ethers are typically better donors than ketones or carboxylates, most likely because the oxygen in the former is already involved in two single bonds, whereas the latter is involved in one double bond. Although oxygen

has two "free pairs" of electrons in both cases, the other electrons have more room to spread out when they are concentrated in the double bond rather than dispersed among the two single bonds. This is critical in any hydrogen bonding, including protonic bridging. A covalent bond to hydrogen is sometimes usefully thought of as a coordination bond because a proton is attached to a donor in a protonic bridge, which is similar to a covalent bond to hydrogen in that it attracts a partially positive hydrogen atom rather than a bare proton. The availability of the electron pair in a concentrated form is important in determining the acidity of the hydrogen in the first case and the strength of the protonic bridge in the second. Because a poor donor to the proton allows for more extensive dissociation in a solvent that is a competitive donor, the more acidic the hydrogen is in the first case, the less available the electron pair.

For example, HCl is more acidic than HF, most likely because the electron pairs of the larger chloride ion are too dispersed to function as well as the more concentrated electron pairs of the smaller fluoride ion. Molecular HF is strongly associated with protonic bridging, whereas HCl is not, for the same reason. Water and nitrate ion are two oxygen-containing examples. The former has a charge of – 0.25, whereas the later has a charge of –0.29. Regardless, water is a far better donor than nitrate ion. The difference could be explained by the latter's multiple bonding nature, which allows the "donor" pain molecules that are unused in nitrogen bonding to spread out more and become less accessible to an external acceptor (Fleming, 2011).

CHAPTER

3 Balancing Chemical Reactions

Contents

3.1. INTRODUCTION

To balance chemical equations is a useful skill in chemistry. Chemical reactions are based on the idea that atoms and molecules frequently combine to form new molecules. Chemical reactions begin with reactants (the substances introduced) and end with products. Some chemical reactions you may be familiar with include rusting, which occurs when iron reacts with oxygen and water, and soda losing its bubbles, which occurs when carbonic acid splits into CO_2 and water.

The fundamental idea behind balancing equations is the law of conservation of mass, which states that physical substances such as atoms and molecules cannot be created or destroyed. This means that both sides of a chemical equation must have the same number of atoms and the same mass of atoms. Take, for example, the simple chemical reaction Ca + Cl_2 gives $CaCl_2$. The equation is already balanced because both sides have an equal number of Ca and Cl atoms. An equation can be balanced by changing the coefficients (numbers placed in front of reactants or products to multiply them) (Laino, 2018).

A subscript, which appears in a smaller print to the right of a molecule, differs from a coefficient, which appears to the left. The coefficient represents the number of molecules. The subscript indicates the number of atoms of a specific element in each molecule. For example, in $3O_2$, the coefficient and subscript are 3 and 2, respectively. To calculate the total amount of the atoms present, multiply the coefficient by the subscript, or the total number of molecules times the average number of atoms in each molecule. $3O_2$ is composed of 6 O atoms. When balancing equations, the subscript of a molecule is never changed; only the coefficient is changed. Changing the subscript would change the chemical make-up of the molecule rather than changing the number of molecules (Klippenstein, 2013).

A balanced chemical equation has an equal number of atoms on both the reactant and product sides for each element involved in the reaction. This condition must be met by the equation in order to be consistent with the law of conservation of matter. It is possible to verify it by adding up the atom counts on either side of the arrow and seeing if they add up to the same number.

This method confirms the equation for the reaction of methane and oxygen to produce CO_2 and water, as shown below:

$$CH_4 + 2O_2 \longrightarrow CO_2 + 2H_2O$$

A balanced chemical equation can frequently be created from a qualitative description of a specific chemical reaction using a relatively simple method known as balancing by inspection. Consider the breakdown of water into oxygen and hydrogen molecules. This process is qualitatively represented by an unbalanced chemical equation:

$H_2O \longrightarrow H_2 + O_2$(unbalanced)

The number of H and O atoms on either side of this equation confirms the equation's imbalance:

Element	Reactants	Products	Balanced?
H	$1 \times 2 = 2$	$1 \times 2 = 2$	$2 = 2$, yes
O	$1 \times 1 = 1$	$1 \times 2 = 2$	$1 \neq 2$, no

Element	reactants	products	balanced?
H	$4 \times 1 = 4$	$2 \times 2 = 4$	$4 = 4$, yes
O	$2 \times 2 = 4$	$(1 \times 2) + (2 \times 1) = 4$ $4 = 4$, yes	

The H atom count is the same on both the reactants and product sides of the equation, but the O atom count is not. The coefficients of the equation can be adjusted as needed to achieve balance.

Remember that changing the subscripts in the formula, which help to characterize the substance's identity, will change the qualitative meaning of the equation. For example, changing the formula of the reactant from H_2O to H_2O_2 would result in atomic balance but would also change the reactant's identity (making it hydrogen peroxide instead of water). The O atom balance can be achieved by increasing the coefficient for H_2O to 2 (Laino, 2018).

$2H_2O \longrightarrow H_2 + O_2$ (unbalanced)

This change threw the H atoms out of balance, but it is simple to restore by changing the coefficient for the H_2 product to 2.

$2H_2O \longrightarrow 2H_2 + O_2$(balanced)

Elements	reactants	products	balanced?
O	$2 \times 1 = 2$	$1 \times 2 = 2$ $2 = 2$, yes	

Element	Reactants	Products	Balanced?
H	$2 \times 2 = 4$	$2 \times 2 = 4$ $4 = 4$, yes	
O	$2 \times 1 = 2$	$1 \times 2 = 2$ $2 = 2$, yes	

Because these coefficients produce an equal number of H and O atoms on the reactant and product sides, the balanced equation is as follows:

$2H_2O \longrightarrow 2H_2 + O_2$

Ø ***Example 1***: Dinitrogen pentoxide (N_2O_5) is formed by the reaction of molecular nitrogen (N_2) and oxygen (O_2). Create a balanced equation for this reaction.

Solution First, create the unbalanced equation.

$N_2 + O_2 \longrightarrow N_2O_5$(unbalanced)

Then, figure out how many of each type of atom are in the unbalanced equation.

Element	Reactants	Products	Balanced?
N	$1 \times 2 = 2$	$1 \times 2 = 2$	$2 = 2$, yes
O	$1 \times 2 = 2$	$1 \times 5 = 5$ $2 \neq 5$, no	

Even though nitrogen is balanced, changes in coefficients are required to balance the number of oxygen atoms. In order to balance the number of oxygen atoms, the least common multiple for the O atom subscripts in these two formulas, 10, should be used to convert the coefficients for the O_2 and N_2O_5 to integers (Klippenstein, 2013).

$N_2 + 5O_2 \longrightarrow 2N_2O_5$(unbalanced)

Element	Reactants	Products	Balanced?
N	$1 \times 2 = 2$	$2 \times 2 = 4$	$2 \neq 4$, no
O	$5 \times 2 = 10$	$2 \times 5 = 10$	$10 = 10$, yes

This change has thrown off the balance of N atoms, which is corrected by increasing the coefficient of the reactant N_2 to 2.

$2N_2 + 5O_2 \longrightarrow 2N_2O_5$

Element	Reactants	Products	Balanced?
N	$2 \times 2 = 4$	$2 \times 2 = 4$	$4 = 4$, yes
O	$5 \times 2 = 10$	$2 \times 5 = 10$	$10 = 10$, yes

Because there are now an equal number of N and O atoms on each side, the equation is balanced.

Ø ***Practice Question:*** Create a balanced equation to explain how ammonium nitrate (NH_4NO_3) degrades into molecular nitrogen, oxygen, and water. Balance oxygen last because it can be found in multiple molecules on the right side of the equation (Laino, 2018).

Ø ***Answer:***

$$2NH_4NO_3 \longrightarrow 2N_2 + O_2 + 4H_2O_2$$

When balancing a chemical equation, using fractions as intermediate coefficients rather than integers can be more practical. Once equilibrium is reached, the fractional coefficients in the equation can be converted to integers without disrupting the atom balance by multiplying all of the coefficients by a whole number. Consider the unbalanced equation for ethane (C_2H_6) reacting with oxygen to produce H_2O and CO_2 (Klippenstein, 2013).

$$C_2H_6 + O_2 \longrightarrow H_2O + CO_2 \text{ (unbalanced)}$$

Using the standard inspection method, one could first balance the C and H atoms by adjusting the coefficients for the two product species, as shown.

$$C_2H_6 + O_2 \longrightarrow 3H_2O + 2CO_2 \text{ (unbalanced)}$$

Because an integer coefficient with the O_2 reactant cannot produce the odd number of seven O atoms produced on the product side of the equation, a fractional coefficient, 7, is used to produce the following provisionally balanced equation:

$$C_2H_6 + 7O_2 \longrightarrow 3H_2O + 2CO_2$$

To create a standard balanced equation with only integer coefficients, multiply each coefficient by two:

$$2C_2H_6 + 7O_2 \longrightarrow 6H_2O + 4CO_2$$

Last but not least, when using balanced equations, it is customary to use the smallest whole-number coefficients. Although the equation for the formation of ammonia by the reaction of nitrogen and hydrogen molecules is balanced.

$$3N_2 + 9H_2 \longrightarrow 6NH_3$$

The coefficients do not use the smallest integers to represent the relative numbers of reactant and product molecules. By dividing each coefficient by the greatest common factor, which is 3, the preferred equation is obtained.

$$N_2 + 3H_2 \longrightarrow 2NH_3$$

Water (H_2O) is made up of hydrogen and oxygen. Consider the following scenario: we combine some elemental oxygen (O_2) and hydrogen (H_2) and allow them to react to form water. The process, known as a chemical reaction, can be summarized as "hydrogen and oxygen react to make water."

The process of water formation. When hydrogen and oxygen combine, they form water. The hydrogen gas from the SS Hindenburg reacts with the oxygen in the air to form water (Klippenstein, 2013).

To make writing easier, we use formulas rather than names when describing a reaction. Symbols can also be used to represent additional words in the reaction. An arrow (→) represents the chemical change, and a plus sign (+) connects the starting and ending substances (if there is more than one).

$$H_2 + O_2 \rightarrow H_2O$$

A chemical equation is simplified way to represent a chemical change, such as this statement. The substances on the left side of the arrow are reactants, and the substances on the right side of the arrow are products. Each formula frequently includes the phases (s) for solid, (l) for liquid, (g) for gas, and (aq) for a substance dissolved in water, also known as an aqueous solution. If phase labels for the reactants and products were included, the reaction would take the following form under typical environmental conditions:

$$H_2(g) + O_2(g) \rightarrow H_2O(l)$$

Chemical equations can also be used to describe physical changes. This will be seen in the near future. This equation is still incomplete because it does not follow the law of conservation of matter. Take a note of how many atoms of each element are present on each side of the arrow. On the reactant side, there are two hydrogen atoms and two oxygen atoms, but only one oxygen atom on the product side. The imbalance is caused by the difference in the number of oxygen atoms on each side of the equation (Laino, 2018).

To make this chemical equation comply with the law of conservation of matter, we must adjust the quantities of the reactants and products as needed to obtain the same number of atoms of a particular element on each side. Because each substance has a unique chemical formula, we are unable to change their chemical formulas. For example, we cannot convert the formula for atomic oxygen to O. However, we can anticipate a range of reactant or product molecule counts. For example, perhaps two water molecules are formed rather than just one:

$$H_2(g) + O_2(g) \rightarrow 2H_2O(\ell)$$

The 2 preceding the water equation is known as a coefficient. It implies the formation of two water molecules. There are now two oxygen atoms on each side of the equation.

This point must be emphasized again because it is critical. You cannot change the formula of a chemical substance to balance a chemical reaction! You must use the correct chemical formula for the substance.

Unfortunately, by including the coefficient 2 at the start of the water formula, we have changed the number of hydrogen atoms on the product side. As a result, we no longer have the same number of hydrogen atoms on both sides. This is easily remedied, by including a coefficient of 2 before the diatomic hydrogen reactant:

$2H_2(g) + O_2(g) \rightarrow 2H_2O(l)$

At this point, each side of the equation has two oxygen atoms and four hydrogen atoms. The reactants and products now have the same number of atoms of each element, satisfying the law of conservation of matter. The reaction is considered balanced. The diatomic oxygen has a coefficient of one in balanced chemical equations, which is usually assumed rather than written (Klippenstein, 2013).

Chemical equations should be balanced. By recording balanced reactions, a chemist can acknowledge the law of conservation of matter.

Ø ***Example 2*****:** Isn't every chemistry equation balanced?

a. $2Na(s) + O_2(g) \rightarrow 2Na_2O(s)$

b. $CH_4(g) + 2O_2(g) \rightarrow CO_2(g) + 2H_2O(\ell)$

c. $AgNO_3(aq) + 2KCl(aq) \rightarrow AgCl(s) + KNO_3(aq)$

Ø ***Solution:***

a. Counting, we discover that the reactants have two sodium atoms and two oxygen atoms, whereas the products have four sodium atoms and two oxygen atoms. This formula is out of balance.

b. There are four hydrogen atoms, four oxygen atoms, and one carbon atom in the reactants. Four hydrogen atoms, four oxygen atoms, and one carbon atom are found in the end products. This calculation is accurate.

c. Two potassium atoms, two chlorine atoms, two oxygen atoms, one nitrogen atom, and one silver atom make up the reactants. The products contain one silver atom, one chlorine atom, one potassium atom, one nitrogen atom, and three oxygen atoms. This equation is unbalanced because there are different numbers of chlorine and potassium atoms

How does one balance a chemical equation using the correct formulas for the reactants and products as a starting point? A back-and-forth strategy is used, with the number of atoms of one element counted on one side, the number of those atoms checked on the other, and a coefficient changed as needed. Check a different element by switching back and forth between one side of the equation and the other when each element has the same number of atoms on both sides of the arrow. In many cases, it makes no difference which element is balanced first and which is balanced last as long as all of the elements have the same number of atoms on each side of the equation (Laino, 2018).

The following rules are for formulating and balancing chemical equations.

i. Construct the correct chemical equations for each reactant and product. Make the fundamental equation.

ii. Count the number of atoms of each element present in both the reactant and the product. A polyatomic ion is considered a unit if it remains unchanged on both sides of the equation.

iii. Individually equilibrate each component by placing coefficients in front of the formulas. When the value is 1, no coefficient is recorded. It is best to begin by balancing components that are unique to one chemical formula on each side of the equation. Subscripts in chemical formulas should never be changed; equations can only be balanced using coefficients (Klippenstein, 2013).

iv. Check that each atom or polyatomic ion has the same value on both sides of the equation.

Reactants:

1C atom 4H ions 2 Cl atom

Products 1C atom 4H ions 8Cl atoms

v. Ensure that each coefficient has the smallest possible ratio. If necessary, reduce to the lowest ratio.

To make the equation equal, for example, write the skeleton equation using the correct formulas as step one.

$CH_4 + Cl_2 \rightarrow CCl_4 + HCl$

Ø **Step 1:** Count the number of polyatomic ions or atoms on each side of the equation.

Ø **Step 2:** We find that a single carbon atom already balances both sides. After that, the hydrogen atoms are balanced. Given that the reactant side contains four hydrogen atoms, the product side must contain the same number. The equation is balanced by inserting a

4 before the HCl:

$CH_4 + Cl_2 \rightarrow CCl_4 + 4HCl$

Reactants:

1C atom 4H ions 2 Cl atom

Products 1C atom 4H ions 8Cl atoms

Each side now has four hydrogen atoms. Because the product side contains eight chlorine atoms (four from CCl_4 and four from the four molecules of HCl), we need eight chlorine atoms as reactants. Because elemental chlorine is a diatomic molecule, four chlorine molecules are required to produce a total of eight chlorine atoms. We add an additional 4: before the Cl_2 reactant.

$CH_4 + 4Cl_2 \rightarrow CCl_4 + 4HCl$

Reactants 1C atom 4H ions 8Cl atoms

Products 1C atom 4H ions 8Cl atom

- Ø **Step 3:** We now check to see if each side has one carbon atom, four hydrogen atoms, and eight chlorine atoms. The chemical equation is balanced. Furthermore, the coefficients have the lowest possible ratio (Klippenstein, 2013).
- Ø ***Example 3:*** Aqueous solutions of lead(II) nitrate and sodium chloride are combined. The reactions' byproducts are an aqueous solution of sodium nitrate and a solid precipitate of lead(II) chloride. Make a balanced chemical equation for this reaction (Laino, 2018).

As part of the first solution, write the skeleton equation using the correct formulas.

$Pb(NO_3)_2(aq) + NaCl(aq) \rightarrow NaNO_3(aq) + PbCl_2(s)Pb(NO_3)_2$

- Ø **Step 1:** Count the number of polyatomic ions or atoms on each side of the equation.

Reactants:

1Pb atom 2 NO^{-3} ions 1Na atom 1Cl atom

Products:

1Pb atom 1NO^{-3} ions 1Na atom 2Cl atoms

- Ø **Step 2:** Resolve the issue. The chlorine atoms and nitrate ions are out of balance. Put a 2 in front of the NaCl at first. As a result, the reactant count has increased by two Na and two Cl atoms. Then add a 2 before the $NaNO_3$. The end result is:

$Pb(NO_3)_2(aq) + 2NaCl(aq) \rightarrow 2NaNO_3(aq) + PbCl_2(s)$

- Ø **Step 3***:* The count for atom and polyatomic ions becomes:

Reactants:

1Pb atom $2NO^{-3}$ ions 2Na atom 2Cl atom

Products:

1Pb atom 2 NO^{-3} ions 2Na atom 2 Cl atoms

- Ø **Step 4: Think About the Outcome:** The equation is now balanced because each element has an equal number of atoms on both sides of the equation. Furthermore, the coefficients have the lowest possible ratio.

Despite the fact that chemical compounds are broken up and new compounds are created during a chemical reaction, atoms in the reactants do not vanish, nor do new atoms suddenly appear to form the products. During chemical reactions, atoms never create new ones or destroy old ones. The atoms in the products are the same as the atoms in the reactants; they have simply been rearranged into different configurations. A complete chemical equation must have the same number of atoms on both the reactant and product sides.

3.2. LAW OF CONSERVATION OF MATTER AND BALANCING CHEMICAL EQUATIONS

The previous section provided a chemical equation for the reaction of hydrogen gas and oxygen gas to produce water in the gaseous phase:

$O_2(g) + H_2(g) \rightarrow H_2O\ (l)$.

On the right side of the equation, there is only one oxygen atom, whereas there are two oxygen atoms on the left. Where did that lone oxygen atom go? Matter cannot just disappear. Again, in chemical reactions, atoms are never created or destroyed (Laino, 2018).

Consider what would happen if we began with two hydrogens and one oxygen molecules. Two water molecules are produced as a result of this combination, with four hydrogen atoms and two oxygen atoms on the left side of the equation and four hydrogen atoms and two oxygen atoms on the right. There has been no violation of the law of conservation of matter. The aforementioned reaction's chemical equation would be:

$2H_2(g) + O_2(g) \rightarrow 2H_2O\ (l)$.

This kind of chemical equation is known as a balanced chemical equation. A chemical equation cannot be considered balanced unless the law of conservation of matter is observed (Klippenstein, 2013).

3.3. COEFFICIENTS AND SUBSCRIPTS

Chemical equations employ two types of numbers. There are subscripts in the chemical formulas of the reactants and products:

$2H_2(g) + O_2(g) \rightarrow 2H_2O(l)$

Once the formulas for the reactants and products are established, the subscripts cannot be changed. Furthermore, before chemical formulas, coefficients are added to show how many molecules of the substance are consumed or generated:

$2H_2(g) + O_2(g) \rightarrow 2H_2O(l)$

By changing the coefficients, the equation can be balanced. Two molecules of water are formed when two molecules of hydrogen and one molecule of oxygen combine, according to the equation above.

Because the identities of the reactants and products are fixed, the equation cannot be balanced by changing the subscripts of the reactants or products. This would change the chemical identity of the species being described. For example, if the equation for the reaction $H_2(g) + O_2(g)$ gives $H_2O(l)$ was balanced by changing the oxygen's subscript on H_2O to H_2O_2, the reaction would no longer produce water. In fact, the equation is correct. It does, however, show how hydrogen peroxide, a completely different substance, is produced:

$H_2(g) + O_2(g) \rightarrow H_2O_2(l)$

Chemical equations are thus rebalanced by changing the coefficients:

$2H_2(g) + O_2(g) \rightarrow 2H_2O(l)$

In chemical formulas, coefficients of "1" are implied in the same way that subscripts of "1" are (the formula for water is written as H_2O, not H_2O_1). In other words, despite the fact that the number "1" is not written, it is the coefficient for $O_2(g)$ in the above balanced equation.

Coefficients in chemical equations can be compared to algebraic coefficients. Coefficients are not written in the middle of a term or variable in algebra, but rather at the beginning. A coefficient in a chemical equation, like in algebra, denotes multiplication. In other words, when we say "$2H_2O$," we're talking about a total of 22, or 4, hydrogen atoms (Klippenstein, 2013).

3.4. ADDITIONAL CHEMICAL EQUATION BALANCING STRATEGIES

"Balancing by inspection," also known as trial and error, is the most fundamental and widely used technique for balancing chemical equations. A few techniques are used to reduce the number of trials and errors required to arrive at an equation that is ultimately correctly balanced:

1. The most complicated formula should usually be balanced first.
2. In general, the simplest formula should be balanced last.
3. The coefficients can be computed by taking the least common multiple of the two numbers.
4. If polyatomic ions are present on both sides of the chemical equation, they can be balanced as a single unit.
5. Fractional coefficients may be required when balancing the equation using the simplest formula left. When a fractional coefficient is used, multiply both sides of the equation by the denominator to get whole-number coefficients.
6. Count the number of atoms of each element on each side of the equation to ensure that the chemical equation is balanced (Laino, 2018).

Ø ***Example 4: The Oxidation of Aluminum:*** Aluminum metal reacts chemically with oxygen gas to produce solid aluminum oxide as follows:

$Al(s) + O_2(g) \rightarrow Al_2O_3(s)$

Ø ***Solution:***

1. The most complicated formula, $Al_2O_3(s)$, should be balanced first, followed by the simplest formula, Al(s).
2. There are two oxygen atoms on the left and three on the right. Six is the least common number between two and three. To create 6 oxygen atoms on each side, a coefficient of "3" is added in front of $O_2(g)$ and a coefficient of "2" is added in front of $Al_2O_3(s)$:

   $Al(s) + 3O_2(g) \rightarrow 2Al_2O_3(s)$

3. Despite this, aluminum atoms remain unbalanced. At this point, 4 aluminum atoms are on the right and 1 aluminum atom is on the left. To balance the aluminum atoms, Al (s) is given a coefficient of "4":

$4Al(s) + 3O_2(g) \rightarrow 2Al_2O_3(s)$

4. Perform one final check to ensure that the equation is still balanced for all elements. When balancing an equation, compiling the data into a table and counting each atom of an element can be useful:

Ø ***Example 5: Potassium Chlorate Decomposition:*** The chemical equation must be balanced when solid potassium chlorate is broken down by heat into solid potassium chloride and oxygen gas. It is possible to see the triangle that appears above the arrow to indicate that heat is required for this reaction to occur:

$KClO_3(s)^{-} \rightarrow KCl(s)+O_2(g)$

Ø ***Solution:***

1. Because $KClO_3(s)$ is the most complicated formula, it should be achieved first. Both KCl (s) and $O_2(g)$ appear to be simple. However, since the potassium and chlorine atoms appear to be in equilibrium, one could begin by balancing the oxygen atoms. Let's keep a running tally this time (Klippenstein, 2013).

2. There are three oxygen atoms on the left and two oxygen atoms on the right. Six is the least common number between two and three. To create 6 oxygen atoms on each side, a coefficient of "2" and a coefficient of "3" are added in front of $KClO_3(s)$ and $O_2(g)$, respectively.

$2KClO_3(s){-}{\rightarrow}KCl(s)+3O_2(g)$

3. As a result, the potassium and chlorine atoms are now out of balance. However, this is easily corrected by adding a "2" coefficient before KCl (s):

$2KClO_3(s){-}{\rightarrow}2KCl(s)+3O_2(g)$

4. Perform final check to ensure that the equation is still balanced for all elements.

Ø ***Example 6: Butane Combustion:*** The following chemical equation must be balanced in order to produce CO_2 and water in the gaseous phase during the combustion of butane liquid (C_4H_{10}) in the presence of oxygen gas:

$C_4H_{10}(l) + O_2(g) \rightarrow CO_2(g) + H_2O(l)$

Ø ***Solution:***

1. The simplest formula, $O_2(g)$, should be balanced last, while the most complex, $C_4H_{10}(l)$, should be balanced first. Let's keep a running tally again, but this time with numbers (Laino, 2018).
2. One carbon atom is on the right and four are on the left. Four is the least common multiple of four and one. $CO_2(g)$ is multiplied by a "4" coefficient to yield four carbon atoms on the right:

   $C_4H_{10}(l) + O_2(g) \rightarrow 4\ CO_2(g) + H_2O(g)$
3. On the right, there are two hydrogen atoms, and 10 on the left. 10 is the least common multiple of 2 and 10. On the right, $H_2O(g)$ is given a coefficient of "5" to produce 10 hydrogen atoms:

   $C_4H_{10}(l) + O_2(g) \rightarrow 4CO_2(g) + 5H_2O(g)$
4. Only oxygen atoms remain unbalanced, as originally intended. At this point, there are two oxygen atoms on the left and 13 on the right. As a result, the equation should be balanced by inserting a coefficient (x) in front of $O_2(g)$:

   $C_4H_{10}(l) + xO_2(g) \rightarrow 4CO_2 + 5H_2O(l)$
5. Given that we must ask ourselves, "What number times two equals 13?" and that we must use a fractional coefficient to balance the oxygen atoms, it appears that Strategy #5 (listed above) is appropriate at this time. In algebra, this problem would be written as $2x = 13$.

By dividing each side by two, one can find a solution for x: x = 13 divide by 2

In other words, the coefficient "13 divided by 2" may come before $O_2(g)$.

$C_4H_{10}(l)+13/2O_2(g)\rightarrow4CO_2(g)+5H_2O(l)$

6. As stated in Strategy #5, when using fractional coefficients, both sides of the equations must be multiplied by the denominator to obtain whole number coefficients (Klippenstein, 2013).

$2\times[C_4H_{10}(l)+13/2O_2(g) \rightarrow 4CO_2(g)+5H_2O(l)$

$2C_4H_{10}(l)+13O_2(g) \rightarrow 8CO_2(g)+10H_2O(l)$

7. Run one last check to ensure that the equation is still balanced across all elements.

Ø ***Example 7: Precipitation of Lead(II) Chloride:*** Aqueous solutions of lead(II) nitrate and sodium chloride are combined. Byproducts of the reaction include lead(II) chloride precipitate and sodium nitrate aqueous solution. Make a balanced chemical equation for this reaction.

Ø ***Solution:***

1. Use the word equation to create a chemical equation. Remember that the nitrate ($NO^{-3}NO^{3-}$) polyatomic ion can be balanced as a unit.

   $Pb(NO_3)_2\ (aq) + NaCl\ (aq) \rightarrow PbCl_2\ (s) + NaNO_3\ (aq)$

2. Adding a "2" in front of $NaNO_3$(aq) appears to simply balance nitrate ions, according to the tallies.

   $Pb(NO_3)_2(aq) + NaCl\ (aq) \rightarrow PbCl_2(s) + 2NaNO_3(aq)$

3. At this point, the sodium and chlorine atoms were out of balance. However, this is easily corrected by adding a "2" coefficient before NaCl (aq):

   $Pb(NO_3)_2(aq)+2NaCl(aq)\rightarrow PbCl_2(s)+2NaNO_3(aq)$

4. A final check shows that the equation is balanced for all elements.

To summarize, in order to be effective, chemical equations must always be balanced. Balanced chemical equations have the same number and type of atoms on both sides. The coefficients of a balanced equation must be the simplest whole number ratio. Chemical reactions are always mass-conserving.

3.5. ATOMS COUNTING IN A FORMULA

Atomic quantification What's the point if one can correctly count atoms? You will be able to compute formula/molecular mass, balance chemical equations, and calculate mols for a specific substance. Perform calculations involving mass-mol particles for a variety of substances.

CH_4 [methane]

This formula contains two types of atoms: hydrogen (H) and carbon (C). Have you noticed the small 4 to the right of the word hydrogen? By looking at that number, you can figure out how many times an atom appears in a formula. At the moment, we have four hydrogens. Notice how there is no small number for carbon at the bottom right? According to this, there is one carbon atom. Only numbers greater than two are printed. There is only one if there is no written number. As a result, the CH_4, 1 C, and 4 H have a total of 5 atoms. The outcomes are as follows:

- # C: 1 atom;
- #H: 4 atoms;
- total # atoms: 5.

NH_4OH [ammonium hydroxide]

This formula contains three different atom types: nitrogen (N), hydrogen (H), and oxygen (O). From left to right, read the sentence. Four H, one O, one N, and one H. Because H appears twice in the formula, we must count the number of H atoms: 4 + 1 = 5 H. Why is NH_4OH formatted this way? Why not write it as NH_5O to make counting the atoms easier? You could, but it's usually written as NH_4OH to distinguish between the NH^{4+} and OH^- ions that make up this ionic compound. Don't worry if you don't understand. I'll go over it in more detail in a later post (Laino, 2018).

Here is the tally:

- # N: 1 atom;
- #H: 5 atoms;
- # O: 1 atom;
- total # atoms: 7.

$(NH_4)_2O$ [ammonium oxide]

When comparing the two formulas, it appears that this one is a little more difficult. What is the purpose of the bracket? It is used to group formulas together for ease of use. Take note of how the bracket covers NH_4? And there's a 2 in the bracket's bottom right corner? As a result, there are two types of NH_4.

Because there are two groups of NH_4, we have a total of two N. Because there is a 2 outside the bracket and a 1 from NH_4, we have $1\times 2 = 2$ N. What about him? We take the following approach: $4 \times 2 = 8$ H. Counting O atoms is simple. We have one O because there is no number at the bottom right of O. So, here's what we have:

- # N: 2 atoms;
- #H: 8 atoms;
- # O: 1 atom;
- total # atoms: 11;
- $3(NH_4)_2O$.

The only difference between this formula and #3 is the presence of a 3 in front of the entire formula. What does that number 3 mean? It means there are 3 of the entire formula if it is put in front of the formula. The equivalent would be to say $(NH_4)_2O$ three times: $(NH_4)_2O$, $(NH_4)_2O$, $(NH_4)_2O$ (Klippenstein, 2013).

How many N do we have then, if that is the case? We have 1× 2 ×3 N, which is 6. I wonder how I got that. We have 1 N inside the bracket, then. Then, outside the bracket, there is a 2, indicating that we have two of whatever is inside the bracket. That implies 1×2 (as was the case, for example, #3 above). There is a 3 in front of the entire formula, so we aren't quite finished yet. In other words, whatever we have so far counted for N, there are three times of that (1×2) + (1×2) + (1×2), or 1×2×3. Being alright so far?

Let's move on to H. It resembles N in some ways. 4 × 2 ×3 = 24H. In the bracket, there are four Hs. Then, outside the bracket, there is a 2, indicating that we have two of whatever is inside the bracket. Since we discussed that for example #3 above, that means 4×2. Nearly there. Recall the 3 before the entire formula. That indicates that whatever we have so far counted for H is actually three times as many, or (4×2) + (4×2) + (4×2), which is the same as 4×2×3. That is how 24 H came about (Laino, 2018).

We have 1 x 3 = 3 O for O. So, this is what we have:

- # N: 6 atoms;
- #H: 24 atoms;
- # O: 3 atom;
- total # atoms: 33.

Let's take it a step further. What if the formula becomes even more complex?

$FeC_2O_4 \cdot 2H_2O$ [iron(II) oxalate dihydrate]A new ingredient in the mix! There is a dot in the exact center. What does this imply? In our example, the iron(II) oxalate crystals are entrapped by two water molecules. By the way, that was a quick response. Assume we're attempting to count atoms. What does that imply for us? Assume that $2H_2O$ is a constituent of the entire molecule. We can count the atoms in front of the dot (FeC_2O_4) in our sleep at this point. Our atoms are made up of 1 Fe, 2 C, and 4 O. However, you'll notice that O comes after the dot in H_2O.

Then we move on to the most recent material, "$2H_2O$." Remember what to do if the word H_2O is followed by a number, as in Example #4, where

$(NH_4)2O$ is followed by the number 3. It's the same as saying "We have two bottles of water." How many Hydrogens do we have now? We have 2×2 = 4 H because there is a big 2 in front of H_2O and a small 2 at the bottom right of H. The small 2 in the bottom right corner of H indicates that two Hydrogens. Using the same reasoning, we obtain 1×2 = 2 O for the O count in $2H_2O$. However, because there is some O in the formula before the dot, we must add up all of the O atoms. 4 + 2 must equal 6 O. The result is:

- # Fe: 1 atom;
- #C: 2 atoms;
- # O: 6 atoms;
- # H: 4 atoms;
- total # atoms: 13.

5 $K_3[Fe(C_2O_4)_3]\cdot 3H_2O$ [potassium iron(III) oxalate trihydrate]

Would you like to try your hand at this beast mode formula's atom count? (Before you begin, keep in mind that the entire formula begins with a 5.

- # K: 15 atoms;
- # Fe: 5 atom;
- #C: 30 atoms;
- # O: 75 atoms;
- # H: 30 atoms;
- total # atoms: 155.

Let's take a moment to digest the formula. In general, there are 5 $K_3[Fe(C_2O_4)_3].3H_2O$. We have a dot component, $3H_2O$, in above formula. That simply means there are three H_2O molecules within the crystal. We have $Fe(C_2O_4)_3$ in that crystal. That means we have one Fe and three C_2O_4 groups (Klippenstein, 2013).

Starting with K, there are three of them, but because there is a 5 in front of it, the total number of atoms is 3×5 = 15 K.

There is only one Fe. When we multiply by the number 5, we get 1×5 = 5 Fe.

Then, for C, things get a little more complicated. There are two C, but because there is a 3 outside the bracket, it means two and three. Finally, because there is a 5 in front of the entire molecule, the total number of atoms is 2×3×5 = 30 C (Laino, 2018). O is similar to C, but it is a little more complicated because it appears in two parts – before and after the dot. Let's start with $Fe(C_2O_4)_3$ before the dot: There are four O, but because there is a three outside the bracket, 4×3 = 12. In terms of the part after the dot, $3H_2O$: There is one O, but because there is a three in front of H_2O, we have 1×3 = 3. So, we have 12 + 3 = 15 in total. However, because we have a 5 in front of the entire term, we must multiply by 5, 15×5 = 75 O. If you prefer to do it directly, the formula is [(4×3) + (1×3)] 5 = 75 O. .

So, if we can count O, counting H atoms should be a piece of cake. We have 2H because there is a 2 at the bottom of H. However, because there is a 3 in front of H_2O, it is 2×3. Finally, because there is a big 5 in front of the entire term, the total number of H atoms is 2×3×5 = 30 H.

3.6. BALANCING CHEMICAL EQUATIONS PRACTICE EXERCISES

After attempting to balance these 10 equations on your own, check the solutions below. The difficulty level varies, so don't give up if some of them appear too difficult. Just remember to start with the element that appears the smallest and work your way up from there. The best approach to resolving these issues is to proceed slowly and methodically. Trying to absorb everything at once can quickly become overwhelming.

1. $CO_2 + H_2O \rightarrow C_6H_{12}O_6 + O_2$
2. $SiCl_4 + H_2O \rightarrow H_4SiO_4 + HCl$
3. $Al + HCl \rightarrow AlCl_3 + H_2$
4. $Na_2CO_3 + HCl \rightarrow NaCl + H_2O + CO_2$

Ø ***Solutions:***

1. $CO_2 + H_2O \rightarrow C_6H_{12}O_6 + O_2$

The first step in balancing chemical equations is to focus on elements that appear only once on each side of the equation. In this case, both carbon and hydrogen meet this requirement. So, let's start with carbon. There is only one carbon atom on the left, whereas there are six on the right. As a result, we assign a coefficient of six to the molecule on the left that contains carbon.

$6CO_2 + H_2O \rightarrow C_6H_{12}O_6 + O_2$

Let us now look at hydrogen. There are two hydrogen atoms on the left and 12 on the right. So, we'll apply a coefficient of 6 to the hydrogen-containing molecule on the left.

$6CO_2 + 6H_2O \rightarrow C_6H_{12}O_6 + O_2$

Now is the time to check the oxygen levels. There are a total of 18 oxygen molecules (6×2 + 6×1) on the left. On the right, there are eight oxygen molecules. For balancing the right side, we now have two options: A coefficient can be applied to either $C_6H_{12}O_6$ or O_2. If we change $C_6H_{12}O_6$, the coefficients for everything else on the left must also change because the proportion of carbon and hydrogen atoms changes. To avoid this, it is usually best to change the molecule with the fewest elements, in this case, the O_2. As a result, we can multiply the O_2 on the right by a factor of six. Our final response is as follows:

$6CO_2 + 6H_2O \rightarrow C_6H_{12}O_6 + 6O_2$

2. $SiCl_4 + H_2O \rightarrow H_4SiO_4 + HCl$

Because hydrogen appears more than once on the same side of the equation in this situation, we can start with any other element. Let us begin by looking at silicon. Because there is only one silicon atom on each side, we do not need to add coefficients just yet. Let us now look at chlorine. There are four chlorine atoms on the left and one on the right. As a result, a four coefficient will be added to the right (Klippenstein, 2013).

$SiCl_4 + H_2O \rightarrow H_4SiO_4 + 4HCl$

Let us now look at oxygen. Remember that we want to look at all of the components on one side of the equation that appear only once. There is only one oxygen atom on the left, whereas there are four on the right. As a result, we will multiply the coefficient on the left side of the equation by four (Laino, 2018).

$SiCl_4 + 4H_2O \rightarrow H_4SiO_4 + 4HCl$

We're almost finished! All that remains is to count the hydrogen atoms on each side. We're finished because both the left and right sides have eight hydrogens. Our final answer is:

$SiCl_4 + 4H_2O \rightarrow H_4SiO_4 + 4HCl$

Before proceeding, as always, ensure that the number of atoms of each element balances on each side.

3. $Al + HCl \rightarrow AlCl_3 + H_2$

Take care because this is a tricky problem. When one atom is all by itself on either side of the equation, it is easiest to start with an element. To begin, we will count the aluminum atoms on both sides. We don't need to add any more coefficients because we already have one on the left and one on the right. Let us now look at hydrogen. One is to the left as well, but there are two to the right. As a result, we will add a two-coefficient to the left.

$Al + 2HCl \rightarrow AlCl_3 + H_2$

Next, we'll look at chlorine. At the moment, two are on the left and three are on the right. However, it is not as simple as simply adding a coefficient to one side. In order for the number of chlorine atoms on both sides to be equal, we must make two and three equals. The least common multiple will assist us in accomplishing this. In this case, multiplying two by three and three by two yields the lowest common multiple of six. To get the answer, multiply 2HCl by three and AlC13 by two:

$Al + 6HCl \rightarrow 2AlCl_3 + H_2$

We can easily declare that we have completed our work because we have examined every component. However, always double-check. Aluminum is no longer balanced in this case because we added a coefficient to the molecule that contains aluminum on the right side. There is one on the left and two on the right. As a result, we will raise the coefficient by one.

$2Al + 6HCl \rightarrow 2AlCl_3 + H_2$

We have more work to do. A final look at the equation reveals that hydrogen has also gotten out of balance. There are six on the left and only two on the right. After one final tweak, we have our final response:

$2Al + 6HCl \rightarrow 2AlCl_3 + 3H_2$

4. $Na_2CO_3 + HCl \rightarrow NaCl + H_2O + CO_2$

You should be getting the hang of balancing equations by now, and it should be becoming easier. When we look at sodium, we see that it appears only once on the right and twice on the left. As a result, we can add our first coefficient to the NaCl on the right (Laino, 2018).

$Na_2CO_3 + HCl \rightarrow 2NaCl + H_2O + CO_2$

Let's get to carbon now. There are two, one to the left and one to the right, so it is not necessary to add coefficients. We'll save oxygen for last because it can be found in a few different places on the left. Consider hydrogen instead. Because there is one on the left and two on the right, we will add a coefficient to the left.

$Na_2CO_3 + 2HCl \rightarrow 2NaCl + H_2O + CO_2$

When we look at chlorine, we see that it already has two on each side of the balance. Now we can return to the subject of oxygen. Because there are three on the left and three on the right, our final answer is

$Na_2CO_3 + 2HCl \rightarrow 2NaCl + H_2O + CO_2$

CHAPTER

4 Combination Reactions

Contents

4.1. INTRODUCTION

Combination or synthesis reactions are fundamental chemical processes in which two or more substances combine to form a single compound. The reactants merge their constituent atoms or ions in these reactions to create a new product. The product typically exhibits different properties from those of the reactants. The general chemical equation represents combination reactions:

$$A + B \rightarrow AB$$

A and B represent the reactants, and AB represents the product compound formed through the combination reaction (Aspuru-Guzik, 2016) (Figure 4.1).

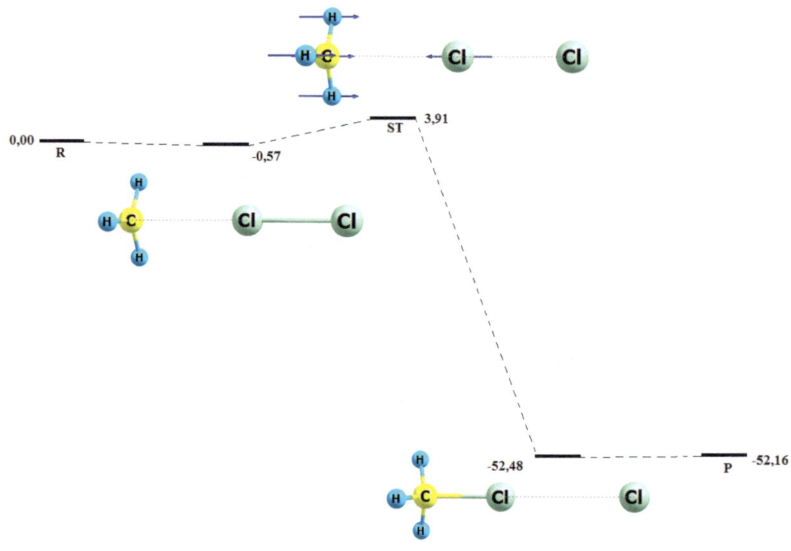

Figure 4.1. *Combination reaction of the methyl radical with the chlorine molecule.*

Source: Resceanu, A., Tran, A.-D., & Magnússon, M., (2020). A comparative study on education for democratic citizenship (EDC) competences in the Icelandic and Romanian context. Social Sciences and Education Research Review. – Scientific Figure on ResearchGate. Available from: https://www.researchgate.net/figure/Combination-reaction-of-the-methyl-radical-with-the-chlorine-molecule_fig8_344173160.

Combination reactions hold great significance in the field of chemistry for several reasons:

1. **Formation of New Substances:** Combination reactions allow the formation of new compounds with distinct properties. These reactions contribute to synthesizing various substances crucial in everyday life, industry, and research.
2. **Chemical Analysis:** Combination reactions often play a vital role in chemical analysis and identification. By combining specific substances and observing the resulting compound, chemists can deduce the presence of particular elements or ions in a sample.
3. **Understanding Chemical Bonding:** Combination reactions provide insights into the nature of chemical bonding. Scientists can better understand the forces that hold molecules together by observing how atoms or ions combine to form compounds (Hu, 2011).
4. **The Basis for Other Reaction Types:** Combination reactions serve as a foundation for other types of chemical reactions. Complex reactions like combustion and precipitation involve a combination of steps in their overall mechanisms.

4.2. UNDERSTANDING CHEMICAL REACTIONS

4.2.1. Basics of Chemical Reactions

Chemical reactions are fundamental processes in which substances transform to form new substances with different properties. These reactions involve the breaking and forming of chemical bonds between atoms or ions. The key aspects of chemical reactions include reactants, products, and the conservation of mass and energy (Aspuru-Guzik, 2016).

1. **Reactants:** These are substances that undergo a chemical change during a reaction. They are written on the left side of a chemical equation.
2. **Products:** These are the new substances formed due to the reaction. They are written on the right side of a chemical equation.
3. **Conservation of Mass:** According to the law of conservation of mass, the total mass of the reactants in a chemical reaction equals the total mass of the products. This means no atoms are created or destroyed during a reaction, only rearranged.
4. **Energy Changes:** Chemical reactions involve the exchange or transfer of energy. Exothermic reactions release energy to the

surroundings, while endothermic reactions absorb energy from the surroundings. (Hu, 2011).

In addition to the aspects above, energy changes are an important component of chemical reactions. Exothermic reactions discharge energy to the environment through heat, light, or sound. Combustion reactions, in which chemicals interact with oxygen to produce heat and light, are examples of exothermic reactions. Conversely, endothermic reactions require energy from the environment to continue. An external energy source, such as heat or electricity, must start these reactions.

Understanding the fundamentals of chemical reactions lays the groundwork for further investigation of the various types of reactions and their unique properties. It enables researchers to create new materials, forecast, and manage chemical reactions, and improve industrial operations. Chemists understand the complex world of chemical transformations and their applications by studying the concepts of reactants, products, conservation of mass, and energy changes (Aspuru-Guzik, 2016).

4.3. FUNDAMENTALS OF COMBINATION REACTIONS

4.3.1. Concept of Reactants and Products

In combination reactions, two or more substances, referred to as reactants, combine to form a single compound, known as the product. The reactants are the starting materials, written on the left side of the chemical equation, while the product is the resulting compound, written on the right side. The general equation $A + B \rightarrow AB$ represents the reaction.

4.3.2. Key Characteristics of Combination Reactions

1. **Reactants and Products:** Various elements, compounds, or ions create brand-new chemicals in combination reactions. The reactants might be substances like sodium (Na) and chlorine (Cl_2) combining to generate sodium chloride (NaCl), or they could be elements like hydrogen (H_2) and oxygen (O_2). A composite made up of merged elements or ions is the end outcome.

   In combination reactions, the reactants combine to create a new molecule with unique features. This procedure includes moving atoms or ions to establish chemical bonds and achieve a more stable structure. A combination reaction produces a substance with unique physical and chemical properties, including its chemical formula, structure, and reactivity.

For instance, the reactants mix to generate water (H_2O) in the interaction between hydrogen gas (H_2) and oxygen gas (O_2). The fact that water is a liquid at room temperature, has a high boiling point and exhibits polarity because of polar covalent connections are only a few of its special qualities. Since these qualities are missing from the individual reactants, combination reactions stand out for their transformative nature (Hu, 2011).

Similar to this, when sodium (Na) and chlorine (Cl_2) mix, sodium chloride (NaCl) is created. This substance has a crystalline structure and high melting and boiling points, and when dissolved in water, it can conduct electricity. These characteristics separate sodium chloride from its parts and show how important combination reactions are in producing new molecules with unique properties.

Understanding the reactants and products in combination reactions is crucial for predicting and studying the behavior of chemical systems. It allows chemists to explore the formation of new compounds, analyze their properties, and further apply this knowledge in fields such as materials science, drug discovery, and environmental chemistry.

2. **Energy Changes:** These occur during combination reactions. Some combination reactions release energy, known as exothermic reactions, while others require an input of energy, known as endothermic reactions. Exothermic reactions release energy through heat, light or sound, while endothermic reactions absorb energy from the surroundings. The energy changes are associated with the breaking and forming of chemical bonds. Energy changes play a significant role in combination reactions. Exothermic reactions release energy due to the formation of new chemical bonds, leading to a more stable product. Usually, this energy is released as heat, light or sound. Fuel combustion and the creation of specific chemicals are two examples (Aspuru-Guzik, 2016).

Conversely, endothermic reactions need an energy input to continue. A drop in temperature or cooling effect results from the energy absorbed by the environment. Endothermic reactions are frequently seen in photosynthesis, where energy from sunlight is absorbed to transform CO_2 and water into glucose, and oxygen photosynthesis, where energy from sunlight is absorbed to transform CO_2 and water into glucose and oxygen; endothermic reactions are frequently seen.

Breaking bonds in the reactants and creating new ones in the product are related to energy changes in combination reactions. While bonds are formed, energy is released, while bonds are broken. Whether a reaction is exothermic or endothermic depends on the overall energy change (Hu, 2011).

3. **Rate of Reaction:** The rate of a combination reaction refers to the speed at which the reactants reacts. Factors such as temperature, concentration of reactants, and presence of catalysts can influence the reaction rate. Higher temperatures generally increase the rate of combination reactions by providing more energy for the particles to collide and react effectively. The rate of a combination reaction is determined by various factors that affect the frequency and effectiveness of collisions between reactant particles. Temperature is critical, as higher temperatures provide greater kinetic energy to the particles, increasing their speed and collision frequency. This leads to more successful collisions and a faster reaction rate.

The concentration of reactants also influences the rate of the combination reaction. Higher reactant concentrations result in more particles per unit volume, increasing the chances of collisions and promoting faster reaction rates.

Catalysts are substances that can significantly enhance the reaction rate without being consumed. They provide an alternative reaction pathway with lower activation energy, allowing more reactant particles to overcome the energy barrier and participate in the reaction. Catalysts can greatly accelerate combination reactions, making them economically and industrially viable.

Other factors, such as the surface area of the reactants and the presence of inhibitors, can also affect the rate of combination reactions. Increasing the surface area exposes more reactant particles, facilitating collisions and promoting faster reaction rates. Inhibitors can slow down or inhibit the reaction by interfering with the mechanism (Aspuru-Guzik, 2016) (Figure 4.2).

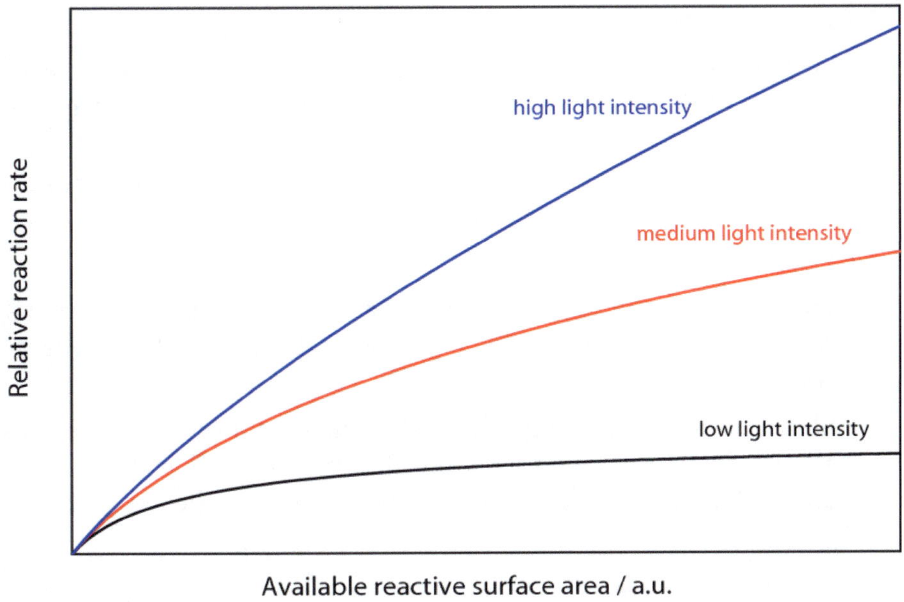

Figure 4.2. *Exemplary illustration of the reaction rate behavior with increased available reactive surface area.*

Source: Bloh, J., (2021). Intensification of heterogeneous photocatalytic reactions without efficiency losses: The importance of surface catalysis. Catalysis Letters. – Scientific Figure on ResearchGate. Available from: https://www.researchgate.net/figure/Exemplary-illustration-of-the-reaction-rate-behavior-with-increased-available-reactive_fig4_349931969.

4.3.3. General Form of Combination Reactions

The equation represents the general form of a combination reaction: $A + B \rightarrow AB$. The reactants A and B combine to form the product AB. The elements or compounds involved in the reaction may vary, but the overall pattern remains the same (Hu, 2011).

Combination reactions can be represented by the general form equation: $A + B \rightarrow AB$, where the reactants A and B combine to form the product AB. While the specific elements or compounds involved may vary, the overall pattern of combining substances remains consistent.

One common example of a combination reaction is the formation of sodium chloride (NaCl) through the combination of sodium (Na) and chlorine (Cl_2):

$2Na + Cl_2 \rightarrow 2NaCl$

In this reaction, two sodium atoms combine with one chlorine molecule to produce two sodium chloride molecules. Sodium chloride is widely used in various applications, including food seasoning, water treatment, and chemical synthesis. Another example is the combination of iron (Fe) and Sulfur (S) to form iron(II)sulfide (FeS):

$Fe + S \rightarrow FeS$

Iron reacts with sulfur to produce iron(II)sulfide in this reaction. Iron(II) sulfide has applications in producing pigments and manufacturing steel and alloys (Aspuru-Guzik, 2016).

Combination reactions are not limited to the combination of elements. They can also involve the combination of compounds. For instance, the combination of calcium oxide (CaO) and CO_2 forms calcium carbonate ($CaCO_3$):

$CaO + CO_2 \rightarrow CaCO_3$

Calcium carbonate is a key component in the formation of shells, coral reefs, and limestone, and it has various industrial uses, such as in the production of cement and as a dietary supplement.

These examples highlight the general form of combination reactions, where different elements or compounds form new substances. Understanding the general form allows scientists to predict and study various combination reactions and their applications in various fields, including chemistry, materials science, and industry (Hu, 2011).

4.3.4. Role of Activation Energy

Activation energy is the minimum energy required to initiate a chemical reaction. In combination reactions, the reactant particles must collide with sufficient energy to break the existing bonds and form new bonds to create the product. The activation energy barrier determines the rate at which the reaction occurs.

The presence of activation energy ensures that only particles with enough energy can overcome the energy barrier and react. Catalysts are substances that lower the activation energy of a reaction, thereby increasing the rate of the combination reaction. They provide an alternative reaction pathway with lower energy requirements, facilitating the formation of products. Understanding the role of activation energy in combination reactions helps scientists to control and manipulate the reaction rates. By adjusting factors such as temperature and the presence of catalysts, they can optimize reaction conditions and improve the efficiency of chemical processes.

The fundamentals of combination reactions involve understanding the concept of reactants and products, the key characteristics such as energy changes and reaction rates, the general form of combination reactions, and the role of activation energy in initiating and controlling these reactions. This knowledge provides a foundation for further exploration of combination reactions and their applications in various scientific and industrial fields.

4.4. COMBINATION OF ELEMENTS

4.4.1. Metal and Non-Metal Combination

Combined reactions between non-metals and metals often result in the creation of ionic compounds. Metal atoms lose electrons to form positively charged ions (cations), while non-metal atoms gain electrons to form negatively charged ions (anions). These oppositely charged ions attract each other to form an ionic compound. For example, the combination of sodium (Na) and chlorine (Cl_2) results in the formation of sodium chloride (NaCl), a common salt (Hu, 2011).

An example is the combination of magnesium (Mg) and oxygen (O_2) to form magnesium oxide (MgO). Magnesium, a metal, loses two electrons to form a magnesium cation (Mg^{2+}), while oxygen, a non-metal, gains two electrons to form an oxide anion (O^{2-}). Combining these ions results in the formation of magnesium oxide, used in producing refractory materials and the medical and agricultural industries.

Metal and non-metal amalgamation reactions are significant in sympathetic to the chemical behavior of elements and the synthesis of various compounds. By controlling the conditions and reactants, scientists can manipulate the outcome of these reactions to get desired products. The ability to predict and switch the formation of ionic compounds through combination reactions has significant implications in materials science, electronics, and various fields of applied chemistry (Aspuru-Guzik, 2016).

4.4.2. Non-Metal and Non-Metal Combination

Combination reactions can also occur between non-metal elements, resulting in the formation of molecular compounds. In contrast to metal and non-metal combinations involving ionic bonding, non-metal, and non-metal combinations involve sharing electrons to form covalent bonds.

These covalent bonds hold the atoms together, resulting in the formation of molecular compounds.

One common example of a non-metal and non-metal combination reaction is the formation of water (H_2O) through the combination of hydrogen (H_2) and oxygen (O_2). In this reaction, two hydrogen atoms share electrons with one oxygen atom, forming covalent bonds. The result is a water molecule where the oxygen atom is bonded to two hydrogen atoms. Water is an essential compound for life, serving as a universal solvent, participating in chemical reactions, and maintaining the temperature balance of organisms.

Another example is the combination of nitrogen (N_2) and hydrogen (H_2) to form ammonia (NH_3). In this reaction, three hydrogen atoms share electrons with one nitrogen atom, forming covalent bonds. The resulting ammonia molecule is one nitrogen atom bonded to three hydrogen atoms. Ammonia has numerous applications, including its use as a fertilizer, cleaning agent, and precursor in producing various chemicals.

Non-metal and non-metal combination reactions synthesize molecular compounds with specific properties and functions. The ability of non-metal elements to share electrons and to form covalent bonds allows for the creation of a wide range of compounds, each with unique characteristics. Understanding and controlling these combination reactions enable scientists to design and synthesize molecular compounds with desired properties for various applications, including pharmaceuticals, polymers, and advanced materials.

4.5. COMBINATION OF COMPOUNDS

4.5.1. Metal Oxide and Acid Combination

When a metal oxide reacts with an acid, salt, and water are produced. The metal oxide provides the metal cation, while the acid provides the anion. For example, the reaction between calcium oxide (CaO) and hydrochloric acid (HCl) results in the formation of calcium chloride ($CaCl_2$) and water (H_2O) (Hu, 2011).

4.5.2. Metal Hydroxide and Acid Combination

Combination reactions between a metal hydroxide and an acid produce salt and water. Metal hydroxides, such as sodium hydroxide (NaOH) or potassium hydroxide (KOH), react with acids to form salts. For instance, combining sodium hydroxide and hydrochloric acid leads to sodium chloride (NaCl) and water formation (Aspuru-Guzik, 2016).

4.5.3. Metal Carbonate and Acid Combination

Combination reactions between metal carbonates and acids produce salt, CO_2, and water. Like calcium carbonate ($CaCO_3$), metal carbonates react with acids to produce CO_2, gas, water, and salt. An example is the combination of calcium carbonate and hydrochloric acid, forming calcium chloride, CO_2, and water.

4.5.4. Metal Bicarbonate and Acid Combination

Like metal carbonates, metal bicarbonates also react with acids to form salt, CO_2, and water. Bicarbonates, such as sodium bicarbonate ($NaHCO_3$), undergo combination reactions with acids to produce salt, CO_2, and water.

4.5.5. Metal Sulfide and Acid Combination

When a metal sulfide reacts with an acid, a salt and hydrogen sulfide gas (H_2S) are formed. Metal sulfides, iron sulfide (FeS) or zinc sulfide (ZnS), react with acids to produce a salt and hydrogen sulfide gas.

4.6. COMBINATION OF METALS WITH NON-METALS

4.6.1. Metal and Halogen Combination

When a metal reacts with a halogen, such as fluorine (F_2), chlorine (Cl_2), bromine (Br_2), or iodine (I_2), a metal halide compound is formed. Metal halides are typically ionic compounds consisting of metal cations and halide anions. These reactions often involve the transfer of electrons from the metal to the halogen. For example, the combination of sodium (Na) with chlorine (Cl_2) produces sodium chloride (NaCl), while the combination of magnesium (Mg) with iodine (I_2) forms magnesium iodide (MgI_2) (Hu, 2011).

4.6.2. Metal and Oxygen Combination

Combination reactions between metals and oxygen result in the formation of metal oxides. Metal oxides can be either ionic or covalent compounds, depending on the nature of the metal and its oxidation state. These reactions are commonly observed in processes such as the rusting of iron. For example, the combination of iron (Fe) with oxygen (O_2) forms iron oxide (Fe_2O_3), which is commonly known as rust.

Understanding the combination reactions between metals and non-metals provides insights into the formation of various compounds and their applications. Metal halides and metal oxides have diverse uses, ranging from industrial catalysts and pigments to materials for electronic devices and construction materials.

Studying the different types of combination reactions expands our understanding of chemical transformations, compound formation, and the interplay between elements and compounds. These reactions are fundamental building blocks in chemistry and play a crucial role in synthesizing new materials, pharmaceuticals, and other chemical products. By examining and manipulating these reactions, scientists can harness their potential to drive advancements in various fields and improve our understanding of the natural world (Aspuru-Guzik, 2016).

4.7. REAL-LIFE APPLICATIONS OF COMBINATION REACTIONS

4.7.1. Combustion of Hydrocarbons

Combustion reactions are crucial in energy production and are widely used in various applications. The combustion of hydrocarbons, which are compounds composed of hydrogen and carbon, is a prime example. Hydrocarbons, such as methane (CH_4), propane (C_3H_8), and octane (C_8H_{18}), undergo combination reactions with oxygen (O_2) in the presence of a spark or heat to produce CO_2 and water (H_2O). This release of energy in the form of heat is harnessed in engines, power plants, and heating system*s* (Figure 4.3).

The combustion of hydrocarbons is a highly exothermic process. When hydrocarbons react with oxygen, the carbon atoms combine with oxygen to form CO_2, while the hydrogen atoms combine with oxygen to form water (H_2O). This reaction releases a significant amount of energy through heat and light. The energy released during combustion can be utilized for various purposes, such as generating electricity in power plants or vehicles (Hu, 2011).

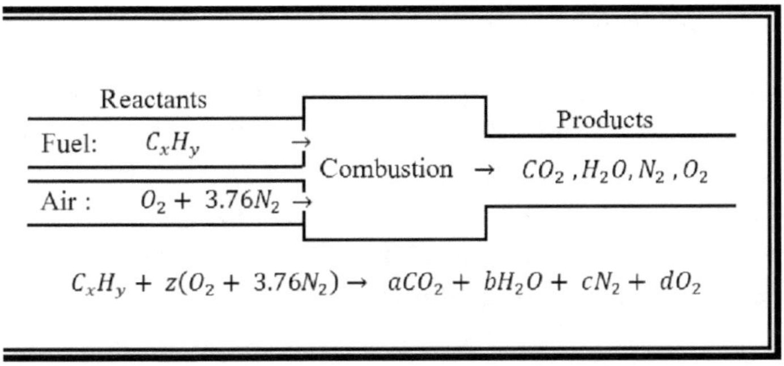

Figure 4.3. *Simplest representation of stoichiometric combustion of hydrocarbon.*

Source: Hossain, A., (2015). Design of a High Intensity Turbulent Combustion System. – Scientific Figure on ResearchGate. Available from: https://www.researchgate.net/figure/15-Simplest-representation-of-stoichiometric-combustion-of-hydrocarbon-fuel_fig10_286451295.

For example, the combustion of methane, the primary component of natural gas, can be represented by the equation:

$$CH_4 + 2O_2 \rightarrow CO_2 + 2H_2O$$

In this reaction, one methane molecule combines with two oxygen molecules to produce one CO_2 molecule and two water molecules. This reaction releases much energy, which can be used to heat homes, cook, or generate electricity.

Similarly, the combustion of other hydrocarbons follows a similar pattern. The combustion of propane, for instance, can be represented by the equation:

$$C_3H_8 + 5O_2 \rightarrow 3CO_2 + 4H_2O$$

The combustion of octane, a component of gasoline, can be represented by the equation:

$$C_8H_{18} + 12.5O_2 \rightarrow 8CO_2 + 9H_2O$$

Combustion reactions of hydrocarbons are essential for meeting our energy needs. However, it is important to ensure proper ventilation and safety measures when dealing with combustion processes to prevent the release of harmful byproducts and minimize the risk of accidents.

4.7.2. Combustion of Fuels

The combustion of fuels, including coal, natural gas, and gasoline, involves combination reactions that release energy. These reactions are vital in transportation, electricity generation, and heating applications. For example, when gasoline (a mixture of hydrocarbons) combines with oxygen during the combustion process in an internal combustion engine, CO_2, water vapor, and energy are produced (Aspuru-Guzik, 2016).

4.7.3. Synthesis of Ammonia

Combination reactions are extensively used in the synthesis of various compounds. One notable example is the Haber-Bosch process, which involves the combination of nitrogen gas (N_2) and hydrogen gas (H_2) to produce ammonia (NH_3). Ammonia synthesis is a vital step in the production of fertilizers, as ammonia serves as a key source of nitrogen for plants.

4.7.4. Synthesis of Water

The synthesis of water (H_2O) through a combination reaction is a fundamental process in both natural and industrial settings. It occurs when hydrogen gas (H_2) combines with oxygen gas (O_2) in the presence of a catalyst or energy source. This reaction is essential for life as water is a universal solvent and it is crucial for many chemical and biological processes (Hu, 2011).

4.7.5. Synthesis of Sulfuric Acid

Another significant application of combination reactions is synthesizing sulfuric acid (H_2SO_4). This compound is widely used in producing fertilizers, detergents, and various chemical processes. Sulfur trioxide (SO_3) combines with water (H_2O) in a combination reaction to form sulfuric acid.

4.7.6. Photosynthesis

Photosynthesis is a vital biological reaction in plants, algae, and some bacteria. It involves the combination of CO_2 and water (H_2O) in the presence of sunlight and chlorophyll to produce glucose ($C_6H_{12}O_6$) and oxygen (O_2). This process is essential for converting solar energy into chemical energy and is the primary source of oxygen in the atmosphere.

4.7.7. Protein Synthesis

Protein synthesis is a complex biological process combining amino acids to form proteins. It occurs in all living organisms and it is crucial for cell and tissue growth, development, and functioning. Combination reactions, specifically the formation of peptide bonds between amino acids are integral steps in protein synthesis.

Understanding the real-life applications of combination reactions provides insights into their significance in various fields. From energy production and chemical synthesis to biological processes, these reactions play a fundamental role in shaping our world and advancing scientific and technological advancements. By harnessing the principles of combination reactions, scientists, and engineers can develop innovative solutions to meet the needs of society and drive progress in industries (Aspuru-Guzik, 2016).

4.8. BALANCING AND PREDICTING COMBINATION REACTIONS

4.8.1. Balancing Combination Reactions

Balancing a chemical equation is essential to ensure that the law of conservation of mass is upheld. When balancing a combination reaction, the goal is to have the same number of atoms of each element on both sides of the equation. The following steps can be followed to balance a combination reaction:

1. **Identify the Reactants and Products:** Determine the elements and compounds involved in the reaction.
2. **Count the Number of Atoms:** Count the number of atoms of each element on both sides of the equation.
3. **Use Coefficients:** Adjust the coefficients before the compounds to balance the number of atoms.
4. **Check the Balance:** Verify that the number of atoms of each element is the same on both sides of the equation.

4.8.1.1. Examples of Balancing Combination Reactions

Ø **Example 1:** Hydrogen gas (H_2) combines with oxygen gas (O_2) to form water (H_2O).

- The unbalanced equation: $H_2 + O_2 \rightarrow H_2O$
- Balancing the equation: $2H_2 + O_2 \rightarrow 2H_2O$

Ø **Example 2:** Sodium (Na) reacts with chlorine gas (Cl_2) to form sodium chloride (NaCl).

- The unbalanced equation: $Na + Cl_2 \rightarrow NaCl$
- Balancing the equation: $2Na + Cl_2 \rightarrow 2NaCl$

Several factors influence the products formed in a combination reaction:

1. **Reactivity:** The reactivity of the elements or compounds involved determines the likelihood of combination. More reactive elements or compounds are more likely to form products.
2. **Stability:** The stability of the resulting compound is also a factor. Some combinations are favored due to the stability of the product compound (Hu, 2011).
3. **Molecular Structure:** The molecular structure of the reactants can affect the type of combination reaction and the resulting products.

4.8.1.2. Examples of Predicting Products

Ø **Example 1:** Combination of magnesium (Mg) and oxygen (O_2):

When magnesium combines with oxygen, it forms magnesium oxide (MgO). The balanced equation is:

$2Mg + O_2 \rightarrow 2MgO$

Ø **Example 2:** Combination of carbon (C) and oxygen (O_2):

When carbon combines with oxygen, it forms CO_2. The balanced equation is:

$C + O_2 \rightarrow CO_2$

Ø **Example 3:** Combination of iron (Fe) and sulfur (S):

When iron combines with sulfur, it forms iron sulfide (FeS). The balanced equation is:

$8Fe + S_8 \rightarrow 8FeS$

By understanding the principles of balancing and predicting combination reactions, chemists can accurately represent chemical transformations and predict the outcomes of reactions. Balancing ensures that the equation obeys the law of conservation of mass while predicting products allows scientists to anticipate the formation of new compounds based on the reactants involved.

These skills are crucial for accurate documentation, experimental design, and the advancement of chemical knowledge (Aspuru-Guzik, 2016).

4.9. CHEMICAL KINETICS IN COMBINATION REACTIONS

4.9.1. Rate of a Reaction in Combination Reactions

The rate of a chemical reaction refers to how quickly reactants are converted into products. In combination reactions, the reaction rate is determined by the speed at which the reactants come into contact and effectively combine to form the desired product. The reaction rate is typically measured by the change in concentration of the reactants or products over time.

Several factors influence the rate of combination reactions:

1. **Nature of the Reactants:** The chemical nature of the reactants plays a significant role in determining the rate of combination reactions. Reactants with higher reactivity and greater affinity for each other are more likely to react quickly. For example, combination reactions between highly reactive metals and non-metals occur rapidly.
2. **Concentration:** The concentration of the reactants affects the rate of combination reactions. Higher concentrations of reactants provide a greater number of particles available for collision, leading to more frequent and effective collisions. This results in an increased rate of reaction.
3. **Temperature:** It has a significant impact on the rate of combination reactions. As temperature increases, the kinetic energy of the reactant particles also increases. This leads to more frequent and energetic collisions, enhancing the reaction rate (Hu, 2011).
4. **Surface Area:** In cases where reactants are in solid form, increasing the surface area of the reactants can accelerate the rate of combination reactions. This is because a larger surface area provides more sites for collisions, leading to more effective collisions and faster reaction rates (Figure 4.4).

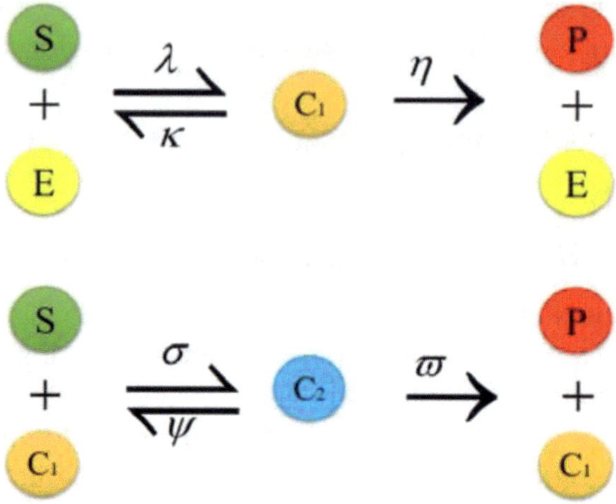

Figure 4.4. *Flowchart of the chemical kinetics.*

Source: Ahmad, Z., Ali, F., Alqahtani, A., Khan, N., & Khan, I., (2021). Dynamics of cooperative reactions based on chemical kinetics with reaction speed: A comparative analysis with singular and nonsingular kernels. Fractals. – Scientific Figure on ResearchGate. Available from: https://www.researchgate.net/figure/Flowchart-of-the-chemical-kinetics_fig1_357018218.

4.9.2. Catalysts and Combination Reactions

Catalysts are substances that can speed up the rate of a chemical reaction without consuming in the process. In combination reactions, catalysts can enhance the reaction rate by providing an alternative pathway with lower activation energy. They achieve this by reducing the energy barrier for the reaction, allowing the reactants to overcome it more easily and form products (Aspuru-Guzik, 2016).

Catalysts work by adsorbing reactant particles onto their surface, facilitating the breaking and formation of chemical bonds. They provide an environment that promotes the reaction and then releases the products, allowing the catalyst to participate in subsequent reaction cycles. Catalysts can significantly increase the rate of combination reactions and are widely used in various industries.

For example, the reaction between nitrogen (N_2) and hydrogen (H_2) to form ammonia (NH_3) in the Haber-Bosch process is catalyzed by iron. Iron as a catalyst increases the reaction rate, enabling the efficient production of ammonia on an industrial scale.

The rate of combination reactions is influenced by factors such as the nature of the reactants, concentration, temperature, and surface area. Catalysts can enhance the reaction rate by providing an alternative reaction pathway with lower activation energy. Understanding the kinetics of combination reactions and the factors affecting their rate is essential for designing and optimizing chemical processes in various fields, including industry, research, and everyday applications (Hu, 2011).

4.10. SAFETY CONSIDERATIONS IN COMBINATION REACTIONS

4.10.1. Hazards and Risks

Combination reactions, like any chemical reaction, can pose certain hazards and risks if improperly handled. It is crucial to be aware of these potential hazards to ensure the safety of individuals and the surrounding environment. Some common hazards associated with combination reactions include:

1. **Heat and Fire Hazards:** Many combination reactions are exothermic, meaning they release heat during the reaction. This heat can increase temperature, fire, or even explosions. It is important to cautiously handle highly reactive substances and exothermic reactions to prevent thermal hazards.
2. **Toxic or Hazardous Products:** Some combination reactions can produce toxic or hazardous substances as products. These products may be harmful if inhaled, ingested, or come into contact with the skin. Proper handling, containment, and disposal of such substances are essential to minimize risks (Aspuru-Guzik, 2016).
3. **Pressure Buildup:** Certain combination reactions may generate gases that can cause pressure buildup in closed systems. If not released or controlled, the pressure can lead to bursting or rupture of containers, resulting in potential injuries or property damage.
4. **Reactivity Hazards:** Combination reactions involving highly reactive substances can exhibit unexpected or uncontrolled reactions. These reactions may result in the release of large amounts of energy, explosions, or the formation of toxic byproducts.

4.10.2. Safety Measures and Precautions

To mitigate the hazards associated with combination reactions and ensure safe handling, several safety measures and precautions should be followed:

1. **Risk Assessment:** Conduct a thorough risk assessment before conducting any combination reaction. Identify potential hazards, evaluate the risks, and implement appropriate control measures to minimize or eliminate those risks.
2. **Proper Training and Knowledge:** Individuals conducting combination reactions should have adequate training and knowledge of the specific reactions and the associated hazards. This includes understanding the properties of the reactants, reaction conditions, and appropriate safety protocols.
3. **Personal Protective Equipment (PPE):** Use appropriate personal protective equipment, such as safety goggles, gloves, lab coats, and closed-toe shoes, to protect against chemical splashes, spills, and other potential hazards.
4. **Ventilation:** Ensure proper ventilation in the laboratory or workspace to prevent the buildup of gases, vapors, or fumes. Adequate ventilation helps maintain a safe working environment by minimizing exposure to hazardous substances (Hu, 2011).
5. **Fire Safety:** Have fire safety measures, including fire extinguishers, fire blankets, and an emergency evacuation plan. Ensure that flammable substances are stored and handled properly, and understand fire suppression techniques clearly.
6. **Chemical Storage and Handling:** Store chemicals appropriately in designated areas, following compatible storage practices. Follow proper handling procedures, including carefully measuring and mixing chemicals, to minimize the risks of combination reactions.
7. **Emergency Preparedness:** Be prepared for emergencies by knowing the location of safety showers, eyewash stations, and emergency exits. Access emergency contact numbers and establish communication protocols in case of accidents or incidents (Aspuru-Guzik, 2016).

Awareness of the hazards and risks associated with combination reactions and implementing proper safety measures and precautions can significantly reduce the potential for accidents or injuries. It is important to prioritize safety and adhere to establish protocols to ensure a safe working environment for everyone involved in combination reactions.

4.11. RECAP OF KEY POINTS

In this comprehensive discussion on combination reactions, we explored the fundamental aspects, types, applications, and safety considerations associated with these reactions. Here is a recap of the key points covered: Combination reactions involve merging two or more substances to form a single compounds. The general equation for a combination reaction is $A + B \rightarrow AB$.

Combination reactions can occur between elements, compounds, or metals with non-metals. Common examples of combination reactions include the combination of metals with non-metals, metal oxides with acids, and synthesizing compounds like water and ammonia.

Combination reactions play a crucial role in various real-life applications, such as combustion, synthesis of compounds, and biological reactions.

Balancing and predicting products in combination reactions require understanding stoichiometry and reaction principles.

The rate of combination reactions can be influenced by factors such as the nature of reactants, concentration, temperature, and the presence of catalysts (Hu, 2011).

Safety considerations are mandatory when dealing with combination reactions due to hazards such as heat, fire, toxic products, and reactivity risks.

4.11.1. Importance and Applications of Combination Reactions

Combination reactions are important in chemistry and have many applications. By understanding and harnessing the principles of combination reactions, scientists, and engineers can achieve the following:

1. **Synthesis of Compounds:** Combination reactions are extensively used in synthesizing various compounds. They allow the formation of complex molecules, such as pharmaceuticals, polymers, and organic compounds (Aspuru-Guzik, 2016).
2. **Energy Production:** Combustion reactions, a type of combination reaction, serve as a vital source of energy in various processes, including burning of fuels for transportation, heating, and electricity generation.

3. **Chemical Industry:** Combination reactions find applications in the chemical industry for producing fertilizers, acids, salts, and other essential chemical compounds.
4. **Biological Reactions:** Combination reactions involve critical biological processes like photosynthesis, protein synthesis, and various metabolic pathways.
5. **Materials Science:** Combination reactions contribute to the synthesis and fabrication of materials with desired properties, such as ceramics, alloys, and composites.

4.11.2. Future Prospects and Research Areas

Combination reactions continue to offer avenues for further exploration and research. Some potential prospects and research areas include:

1. **Catalyst Development:** Advancements in catalyst design and discovery can enhance the efficiency and selectivity of combination reactions. Developing novel catalysts can enable the synthesis of specific compounds with improved yields and reduced energy requirements (Hu, 2011).
2. **Sustainable Chemistry:** Exploring combination reactions focusing on sustainability can lead to developing greener synthesis routes, using renewable feedstock, minimizing waste generation, and reducing impacts of environmental.
3. **Reaction Kinetics:** Further understanding the kinetics of combination reactions, including reaction mechanisms and rate-determining steps can contribute to better reaction optimization and control.
4. **Computational Modeling:** Utilizing computational methods and simulations can aid in predicting and understanding the outcomes of complex combination reactions, enabling more efficient experimental design and reducing trial-and-error approaches (Aspuru-Guzik, 2016).
5. **Application Expansion:** Exploring new applications of combination reactions, such as in energy storage, nanomaterials synthesis, and sustainable agriculture, can lead to innovative solutions and advancements in various industries.

These fundamental processes in chemistry play a vital role in synthesizing compounds, energy production, and biological reactions. Understanding the principles, types, and applications of combination reactions allows us to harness their potential for various purposes. By prioritizing safety, conducting further research, and embracing sustainable practices, the field of combination reactions can continue contributing to advancements in science, technology, and society (Aspuru-Guzik, 2016).

CHAPTER

5 Decomposition Reactions

Contents

5.1. INTRODUCTION

The numerous modifications we notice in the environment around us are driven by chemical reactions. Decomposition reactions stand out among the diverse range of reactions that take place. The breakdown of a compound into simpler molecules occurs during these events, often referred to as analytical reactions, under particular circumstances. Complex molecules are dismantled through the process of decomposition, revealing the mysteries held inside their chemical bonds (Lucht, 2017).

In addition to being crucial to our comprehension of chemistry, decomposition processes also have broad ramifications in several disciplines, including biology, business, and environmental science. Scientists have made significant advancements in a variety of scientific fields by deciphering the complexities of decomposition reactions and gaining priceless insights into the behavior of substances.

Decomposition reactions are fundamentally the reverse of synthesis reactions, which result in the formation of compounds from simpler materials. A single compound is divided into two or more simpler compounds during breakdown process. Chemical bonds within the compound are broken and reorganized during this process, resulting in the creation of several products (Chen, 2010) (Figure 5.1).

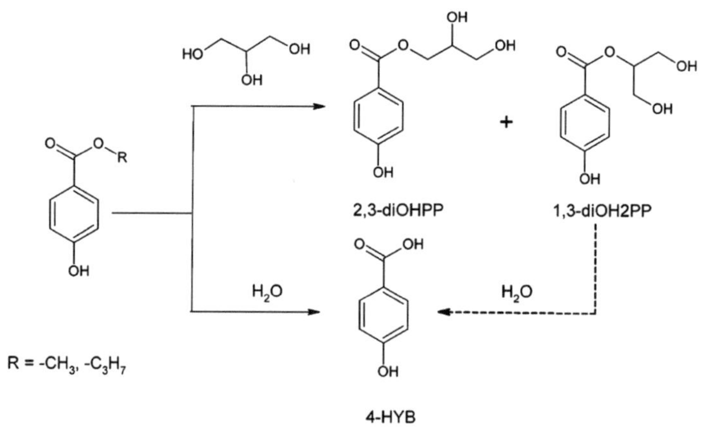

Figure 5.1. *Chemical bonds.*

Source: Farsa, O., Šubert, J., & Marečková, M., (2011). Hydrolysis and trans-esterification of parabens in an aqueous solution in the presence of glycerol and boric acid. Journal of Excipients and Food Chemicals, 2, 41–49. – Scientific

Figure on ResearchGate. Available from: https://www.researchgate.net/figure/Possible-decomposition-reactions-of-MP-and-PP-in-aqueous-solution-in-the-presence-of-GLY_fig2_286615907.

It is essential to appreciate the many types of decomposition reactions as well as the variables that affect their occurrence in order to fully grasp mechanisms at work. Scientists can learn more about the complexities of chemical reactions and the kinetics of molecular interactions by investigating these processes.

Exploring decomposition reactions' fundamental ideas and traits is crucial for understanding how they work. A decomposition reaction is fundamentally the opposite of a synthesis reaction. While decomposition processes break down complex molecules to create two or more simpler substances, synthesis reactions combine simpler components to create more complex compounds (Lucht, 2017).

Understanding many forms in which decomposition reactions take place is one of the fundamental components of the process. The nature of the chemical and the circumstances of the reaction are two elements that affect the sort of breakdown reaction that happens. Thermal, electrolytic, and photolytic decomposition are the three primary categories of decomposition reactions.

1. **Thermal Decomposition:** When a compound is heated to a given temperature, thermal decomposition reactions take place, which lead to the compound disintegrating into more basic elements. Metal carbonates, metal hydroxides, metal nitrates, and metal sulfates are the most prevalent materials that exhibit this kind of disintegration. For instance, when calcium carbonate ($CaCO_3$) is heated, calcium oxide (CaO), as well as CO_2, are produced (Chen, 2010).
2. **Electrolytic Decomposition:** In electrolytic decomposition reactions, a chemical is broken down into its component parts by an electric current. A common example of a substance that experiences electrolytic breakdown is water (H_2O). Water splits into hydrogen gas (H_2) and oxygen gas (O_2) when an electric current is applied to it.
3. **Photolytic Breakdown:** A chemical complex breaks down into simpler components when it is exposed to light energy through a process known as photolytic breakdown. For instance, when chlorine (Cl_2) is exposed to ultraviolet (UV) radiation, it photo

decomposes, producing chlorine radicals (Cl·). Understanding the factors that influence decomposition reactions is crucial for anticipating and controlling these processes. Numerous factors, including temperature, catalysts, pressure, light, concentration, and surface area, can influence the rate and amount of decomposition. These factors determine the amount of energy required to break the chemical bonds inside the compound and have an impact on the reaction's overall kinetics (Lucht, 2017).

Additional information about the quantities of reactants and products involved is provided by the stoichiometry of decomposition processes. Understanding the conservation of mass throughout the reaction requires the balancing of decomposition equations. It enables us to calculate the molar ratios and foretell the products that will be produced. Decomposition reactions have several practical applications in various industries, the environment, and even life. Decomposition processes are used in industry to make fertilizer, remove metals from ores, and produce oxygen. Decomposition processes are used in environmental applications to clean water and reduce air pollution. Decomposition reactions are essential in biology and medicine for the breakdown of organic matter and the degradation of drug inside the body.

We can see the practical applications and significance of decomposition reactions in our daily lives by studying real-world examples and case studies, such as the decomposition of calcium carbonate in limestone, the electrolysis of water for the production of hydrogen, and the photodecomposition of ozone in the ozone layer. It is crucial to remember that decomposition reactions might be hazardous and risky. Some reactions can produce high temperatures or the discharge of hazardous gases. When handling decomposition reactions, it is essential to use caution, adhere to safety procedures, and wear the proper protective equipment. (Chen, 2010).

An essential component of chemical research, decomposition reactions provide a greater knowledge of the behavior of matter and the complex dynamics of chemical changes. We may use this knowledge to improve industrial processes, expand scientific knowledge, and handle environmental issues by discovering the mechanisms underlying these responses. We may fully comprehend the significance of decomposition reactions in forming our universe with a thorough investigation of these processes.

In chemical research, breakdown processes are a key concept that reveal molecules' complexity. By disintegrating these substances, we learn important lessons about behavior of matter behaves and can use decomposition's power in a variety of ways. We can understand the importance of decomposition reactions in expanding scientific understanding, enhancing industrial processes, and addressing environmental issues by thoroughly examining them.

5.2. THE BASICS OF DECOMPOSITION

Decomposition reactions occupy a key position in the broad world of chemical reactions. The breakdown of a compound into simpler molecules occurs during these events, commonly referred to as analytical reactions. It is essential to know the fundamentals of decomposition reactions in order to fully appreciate the underlying concepts of chemical transformations and how they affect many scientific fields (Lucht, 2017).

A decomposition reaction is fundamentally the polar opposite of a synthesis reaction. While decomposition processes break down complex molecules to create two or more simpler substances, synthesis reactions combine simpler components to create more complex compounds. Scientists have learned a great deal about the dynamics of chemical breakdown and the behavior of matter by solving the riddles surrounding decomposition processes.

Understanding the definition, kinds, and stoichiometry involved in balancing the reaction equations of decomposition reactions is crucial for understanding their principles. We can discover the mysteries of chemical breakdown and see its importance in daily life by comprehending these crucial elements.

Understanding the amounts of reactants and products involved in decomposition processes depends heavily on their stoichiometry. By balancing the decomposition equations, we can calculate the molar ratios and forecast the products that will be produced. Scientists can guarantee the conservation of mass during the reaction by studying stoichiometry (Chen, 2010).

We get a greater grasp of chemical breakdown and its ramifications across numerous scientific domains by deciphering the fundamentals of decomposition reactions. Decomposition reactions offer important insights into the behavior of matter and gave the way for technical breakthroughs and scientific discoveries in fields ranging from industry to biology and environmental science.

5.2.1. Common Types of Decomposition Reactions

The disintegration of a compound into simpler chemicals is a key component of decomposition reactions, a fundamental class of chemical processes. Understanding the behavior of matter and the kinetics of chemical transformations depends heavily on these reactions. Exploring the many types of decomposition processes is crucial in developing a thorough grasp of them. These categories develop as a result of the composition of the compound and the circumstances surrounding the reaction. The various ways that compounds can break down and transform into simpler substances can be discovered by investigating the various breakdown reactions. Understanding these types offers useful insights into the complex processes of chemical deterioration and their significance in various scientific fields (Lucht, 2017).

5.2.1.1. Thermal Decomposition

When a chemical is exposed to heat, a specific form of decomposition reaction called thermal decomposition takes place, which causes the complex to break down into simpler substances. This process reveals the complex pathways of chemical deterioration in response to temperature. Scientists have obtained important insights into the behavior of compounds and the transforming force of heat by comprehending the principles and patterns behind thermal decomposition (Chen, 2010) (Figure 5.2).

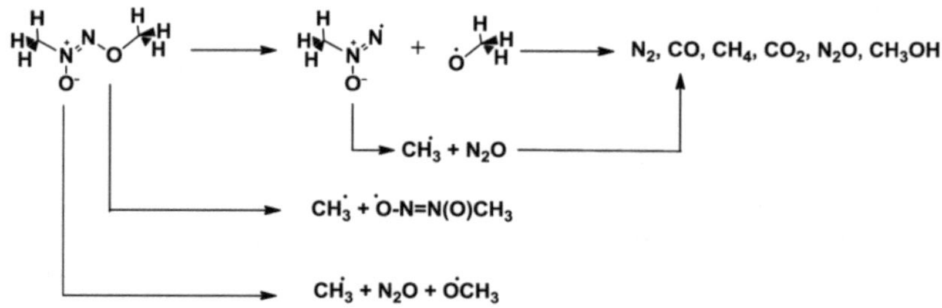

Figure 5.2. *Supposed thermal decomposition scheme presented in the work of N-methyl-N'-methoxydiazene-N-oxide.*

Source: Nikolaeva, E., Khrapkovskii, G., Aristov, I., Egorov, D., & Shamov, A., (2021). Alternative mechanisms of the primary act of gas-phase monomolecular thermal decomposition of N-methyl-N'-methoxydiazene-N-oxide according to the data of quantum chemical calculations. Journal of Physics: Conference Series. – Scientific Figure on ResearchGate. Available from: https://www.researchgate.net/figure/Supposed-thermal-decomposition-scheme-presented-in-the-work-of_fig1_356181899.

The increased kinetic energy of molecules caused by a rise in temperature drives the thermal breakdown processes. The chemical bonds inside the complex are disrupted by this increased energy, which causes the bonds to cleave and give rise to new compounds or elements. The structure and chemistry of the original compound have an impact on the specific products that are produced during thermal decomposition.

Many different compounds, including metal carbonates, metal hydroxides, metal nitrates, and metal sulfates, are prone to thermal breakdown processes. Because each of these substances displays unique thermal degradation traits, it is possible to comprehend the underlying chemical processes better. When heated, metal carbonates, like calcium carbonate ($CaCO_3$), rapidly go through thermal breakdown. Calcium carbonate breaks down into calcium oxide (CaO) and CO_2 when the temperature rises. This reaction is frequently seen in natural limestone, which, when heated, decomposes into lime (calcium oxide) and emits CO_2 (Lucht, 2017).

When heated, metal hydroxides such as copper(II) hydroxide ($Cu(OH)_2$) also decompose thermally. As the temperature rises, the complex decomposes into water (H_2O) and metal oxide (CuO). The synthesis of metal oxides for catalytic uses is one industrial process where this reaction is important. Lead(II) nitrate ($Pb(NO_3)_2$) and other metal nitrates are thermally decomposed to produce nitrogen dioxide (NO_2), oxygen gas (O_2), and the appropriate metal oxide. This reaction illustrates the heat breakdown of metal oxides and the production of oxygen gas.

When heated, metal sulfates, such copper(II) sulfate ($CuSO_4$), show features of thermal breakdown. Sulfur dioxide (SO_2), copper (II) oxide (CuO), and oxygen gas (O_2) are the products of the breakdown of copper (II) sulfate. The manufacture of copper oxide for various uses and the extraction of copper from its ores depend on this reaction. It is significant to remember that the circumstances of thermal breakdown might affect the speed and scope of the reaction. The reaction's kinetics can be altered by elements including pressure, temperature, and the presence of catalysts. For instance, while the addition of catalysts can speed up reactions and alter the products produced, greater temperatures typically speed up thermal degradation (Chen, 2010).

Understanding thermal breakdown reactions offers important insights into how chemicals behave at various temperatures. These reactions have substantial practical ramifications in addition to improving our understanding of chemical dynamics. Controlled thermal decomposition is frequently used in industrial operations to obtain desired materials, including the extraction of metals from ores, the creation of metal oxides, and the synthesis of fertilizers.

A fascinating form of decomposition reaction influenced by heat is thermal decomposition. Scientists can understand the complexities of chemical deterioration and foresee the production of simpler substances by submitting compounds to particular temperature settings. In addition to improving our knowledge of how matter behaves, the study of thermal breakdown has several industrial uses. The mysteries of thermal breakdown are continually being revealed by academics through inquiry and analysis, paving the path for developments across a range of scientific fields.

5.2.1.2. Electrolytic Decomposition

When a compound is exposed to an electric current, a sort of decomposition reaction called electrolytic decomposition takes place that causes the compound to break down into simpler components. This amazing technique demonstrates how electricity may dramatically alter chemical reactions and offers insightful information on how compounds behave when exposed to an electric field.

The principles of electrolysis, which include the conduction of electricity through an electrolyte—a substance capable of carrying electricity when dissolved or molten—are the foundation for the electrolytic breakdown process. The component that needs to be broken down is placed in a suitable medium, frequently an aqueous solution or a molten state, and electrolysis takes place within an electrochemical cell.

A sequence of redox reactions happen as the electrolyte is passed through by the electric current during electrolytic breakdown. These processes take place at the anode and cathode electrodes, where ion migration and electron transfer cause the chemical to break down. Oxidation occurs at the anode, which causes an electron loss. Here, the compound's negatively charged ions or molecules undergo oxidation to produce products. At the cathode, reduction takes place concurrently, resulting in an electron gain. Ions or molecules with a positive charge go in the direction of the cathode, where they undergo reduction to produce various chemicals (Chen, 2010).

The type of substance and the ions in the electrolyte have an impact on the specific products generated during electrolytic decomposition. For instance, electrolytic breakdown takes place when water (H_2O) is exposed to an electric current. Hydroxide ions (OH^-) or water molecules are oxidized at the anode to form oxygen gas (O_2). Positively charged hydrogen ions (H^+) are reduced at the cathode to produce hydrogen gas (H_2) (Lucht, 2017).

In addition to water, other substances can also degrade electrolytically. Among the other compounds that can be electrolyzed are salts and acids. For instance, during the electrolysis of sodium chloride (NaCl), sodium hydroxide (NaOH) is formed at the cathode and chlorine gas (Cl_2) at the anode. Two notable industries that make use of this reaction are the production of chlor-alkali and water treatment.

The settings and parameters of the electrolysis have a significant impact on the rate and degree of decomposition. The strength of the electric current, the amount of electrolyte present, and the design of the electrochemical cells are all factors that affect how well the reaction works. Furthermore, the selection of electrode materials may affect whether or not specific chemical processes are accelerated or prevented.

Significant progress has been made in numerous domains as a result of the study of electrolytic decomposition reactions. The process of electrolysis is used in the manufacture of metals like aluminum and copper, where the compounds are broken down to produce pure metal. In electroplating procedures, where metals are deposited onto surfaces to improve their appearance or qualities, electrolytic breakdown also plays a significant role.

A direct current (DC) power source is necessary for electrolytic decomposition, and safety measures must be taken when working with electricity and potentially dangerous substances. The safe and effective functioning of the process depends on proper insulation, suitable electrode materials, and careful management of the electrolysis conditions.

A fascinating sort of decomposition process is electrolytic decomposition, which uses electric current to accelerate chemical breakdown. Scientists can create simpler substances by electrolyzing complex compounds to reveal the intricate workings of electron transport and ion migration. Our understanding of chemical changes is improved by studying electrolytic breakdown, which also has useful industrial applications. Scientists are exploring the enormous potential of electrolytic decomposition to promote scientific understanding and technological development through innovations (Chen, 2010).

5.2.1.3. Photolytic Decomposition

When a chemical is exposed to light energy, a fascinating form of decomposition reaction called photolytic decomposition takes place that causes the compound to break down into simpler components. This fascinating process shows the transformational power of light and offers important information about how chemicals behave when exposed to particular wavelengths. The absorption of photons—light particles—by the chemical is necessary for photolytic breakdown to occur. When a photon is absorbed, it gives the compound energy, which causes chemical bonds to be broken and new compounds to form. The type of substance and the light wavelengths used determine which specific products are generated during photolytic degradation.

Different substances have varying degrees of photochemical reactivity, and some are extremely sensitive to particular light wavelengths. For instance, photolytic breakdown of chlorine (Cl_2) is known to occur when it is exposed to ultraviolet (UV) radiation. Chlorine molecules absorb ultraviolet light, which causes the Cl-Cl link to break down and produce chlorine radicals (Cl). These extremely reactive radicals are capable of taking part in a wide range of chemical processes, illuminating the transformative potential of light-induced breakdown (Lucht, 2017) (Figure 5.3).

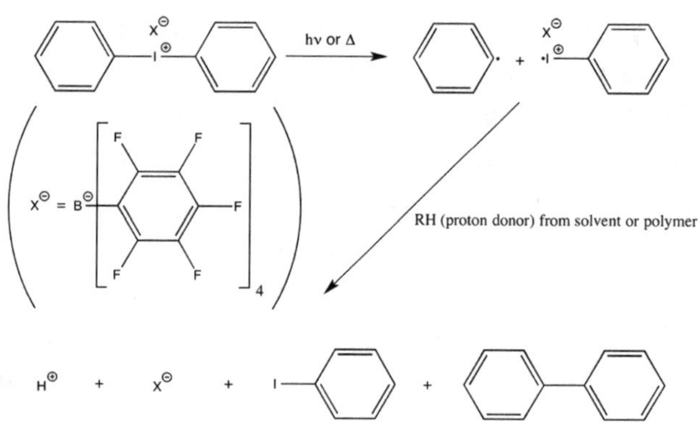

Figure 5.3. *Photolytic/thermolytic decomposition mechanism of diphenyliodonium-type PAGs.*

Source: Uzunlar, E., Schwartz, J., Phillips, O., & Kohl, P., (2016). Decomposable and template polymers: Fundamentals and applications. Journal of Electronic Packaging. – Scientific Figure on ResearchGate. Available from: https://www.researchgate.net/figure/Photolytic-thermolytic-decomposition-mechanism-of-diphenyliodonium-type-PAGs-Adapted_fig8_298334636.

Diatomic molecules like chlorine, also go through photolytic degradation. For instance, when exposed to UV light, some organic molecules, such as halogenated hydrocarbons can degrade. The breaking of carbon-halogen bond brought on by UV photon absorption results in the generation of halogen radicals and reactive carbon fragments. These reactions have an impact on atmospheric chemistry because they help pollutants degrade and ozone-depleting chemicals develop (Chen, 2010).

The intensity and wavelength of the light, the compound's concentration and type, the presence of any sensitizers or catalysts, as well as other variables, affect the efficiency and scope of photolytic decomposition. For a compound to decompose, the energy carried by the photons must be high enough to overcome the bonds' resistance. Consequently, depending on the makeup of their electronic structure and bond energies, compounds may show selective reactivity towards particular light wavelengths.

Application areas f of photolytic breakdown reactions include environmental science, chemical synthesis, and photovoltaics. In environmental science, photodegradation processes that use sunlight are essential for breaking down pollutants and cleaning up toxins in the air, water, and soil. Controlled photolytic breakdown events in chemical synthesis allow for the highly selective generation of particular molecules or intermediates. In photovoltaic devices, where light energy is utilized to drive chemical reactions for the creation of electrical energy, photolytic breakdown is also used.

It is crucial to remember that environmental conditions like temperature, humidity, and the presence of other molecules can affect the effectiveness and selectivity of photolytic breakdown. When working with light sources that emit potentially dangerous radiation, such as UV light, safety procedures must be taken.

The decomposition mechanism known as photolytic decomposition, which takes advantage of light's transforming properties, is fascinating. Scientists can decipher the intricate mechanics of photochemical reactions, which break down compounds into simpler chemicals, by exposing them to various wavelengths of light. In addition to improve our knowledge of light-matter interactions, the research of photolytic decomposition has real-world applications in environmental cleanup, chemical synthesis, and renewable energy. Scientists work to exploit the potential of photolytic decomposition for innovative and sustainable developments through ongoing investigation and study (Lucht, 2017).

5.2.2. Stoichiometry of Decomposition Reactions

Determining the quantities of reactants and products involved in these chemical transformations depends on the stoichiometry of decomposition processes. Scientists can ascertain the molar ratios of substances and forecast the amounts of reactants consumed and products created during a decomposition reaction by balancing the reaction equation and using the concepts of stoichiometry. Before the reaction's stoichiometry can be calculated, a decomposition equation must be balanced. The law of conservation of mass is satisfied when each element has an equal number of atoms on both sides of the equation. This makes it possible to accurately describe the reactants and end products used throughout the breakdown process.

The reactants' and products' coefficients show the relative amounts in which they combine or dissociate after the equation has been balanced. The number of moles, molecules, or particles of each chemical participating in the process can be inferred from these coefficients. Understanding the stoichiometric relationship between the reactants and products can be done by using the balanced equation. The quantity of reactants ingested or products produced in a decomposition reaction can be calculated using the stoichiometry of the reaction. This knowledge is essential for real-world applications like figuring out how much material is needed for industrial processes or forecasting product yields.

One can utilize the idea of molar ratios generated from the balanced equation to compute the amounts of reactants and products. In a chemical equation, the moles of one material to the moles of another substance are represented by a molar ratio. These ratios make it possible to translate the known quantities of one chemical into the amounts of the other material that will be reacting (Chen, 2010).

Take the breakdown of calcium carbonate ($CaCO_3$) into calcium oxide (CaO) and CO_2 as an illustration. The reaction's balanced equation is as follows:

$$CaO(s) + CO_2(g) = CaCO_3(s)$$

One mole of calcium carbonate breaks down into one mole of calcium oxide and one mole of CO_2, as shown by this equation. Calcium carbonate and calcium oxide have a 1:1 molar ratio, and calcium carbonate and CO_2 have a 1:1 molar ratio as well. One can use these molar ratios to determine the corresponding amounts of other substances if they are aware of the amount (in moles or grams) of each reactant and product that participated in the reaction. As a result, it is possible to calculate the masses of the reactants or products, volume ratios in gases, or any other needed number using the reaction's stoichiometry.

The stoichiometry of decomposition processes is crucial for understanding the basic principles underlying chemical transformations as well as for calculating numbers. Understanding the atom rearrangement and mass conservation throughout the reaction is made possible by the balancing equation and stoichiometric relationships. The stoichiometry of decomposition processes necessitates the assumption that the reaction progresses to completion and that all of the reactants are consumed. In reality, though, certain reactions might not finish because of a variety of things, like reversible reactions or the presence of contaminants. When using stoichiometry in practical situations, certain factors should be taken into account (Lucht, 2017).

In conclusion, comprehending the interactions between reactants and products depends on the stoichiometry of decomposition reactions. Scientists can ascertain the quantities of the components involved in the reaction and forecast the amounts of reactants consumed and products created by balancing the reaction equation and using molar ratios. In order to analyze decomposition reactions, stoichiometry provides a fundamental framework. This framework is essential for real-world applications and the expansion of scientific knowledge (Chen, 2010).

5.2.3. Balancing Decomposition Equations

To guarantee the conservation of mass and atom count throughout a chemical transition, it is essential to possess the fundamental chemistry skill of balancing decomposition equations. We can precisely describe the reactants and products used in the decomposition reaction by balancing the equation, which gives us a thorough knowledge of the reaction stoichiometry. A decomposition equation must have an equal number of atoms for each element on both sides in order to be balanced. This is founded on the idea of mass conservation, which holds that during a chemical process, matter cannot be created or destroyed.

We can use the following guidelines to balance a decomposition equation:

1. Determine the species of the reactant and product: Find the molecule that is being broken down (the reactant) and the substances that are produced as a result (the products).

Recognizing the distinct components that make up each chemical is crucial.

2. The imbalanced equation should be written using chemical symbols and formulas, with the reactant on the left-hand side and the products on the right.
3. Start balancing with the species that is the most complex: The elements that are present in the fewest compounds should be balanced first. For instance, balance oxygen (O) last if it only occurs in one combination.
4. Treat polyatomic ions as single units when balancing them: Treat polyatomic ions as single units. This helps balance the charge on both sides of the equation and preserves the integrity of the ion.
5. To balance the number of atoms on each side of the equation, adjust the coefficients. The number of atoms in each compound is multiplied by coefficients before they are placed in front of the formulations.
6. Look over the balance: Verify that there are an equal number of atoms of each element on both sides of the equation by counting them. Adjust the coefficients to bring the equation into balance if necessary.

It is crucial to keep in mind that there are frequently multiple proper ways to balance an equation and therefore balancing decomposition equations may need some trial and error. Additionally, whenever possible, coefficients have to be decreased to their smallest whole-number ratio (Lucht, 2017).

Let's look at an instance to demonstrate how to balance a decomposition equation:

Equation not in balance: AB gives A + B

The compound AB in this instance separates into the components A and B. The atoms that are present in the fewest compounds must first be balanced in order to balance the equation. Assume that A is an element with a single atom and that B is a diatomic molecule.

Equation that balances: $2AB = 2A + B_2$

By adding a coefficient of 2 in front of AB and A in this equation, we have ensured that there are an equal number of atoms on both sides. Additionally, the A atom production is balanced by the coefficient of 2 in front of AB. After balancing, we check to make sure each element has an equal number of atoms on both sides. In this instance, both sides of the equation include two A and two B atoms. In conclusion, correct representation of the reactants and products involved in a chemical transformation depends on the balance of decomposition equations. We guarantee the conservation of mass and atom count by using a methodical approach and modifying factors. Equations can be balanced to help with further analysis and calculations based on the balanced equation and to provide a deeper understanding of the stoichiometry of the reaction (Chen, 2010).

5.3. FACTORS INFLUENCING DECOMPOSITION REACTIONS

Chemical reactions called decomposition reactions cause a molecule to disintegrate into less complex elements. Numerous elements that control the rate, scope, and circumstances of decomposition have an impact on these processes. For both scientific and practical purposes, it is essential to comprehend these elements in order to predict and manage breakdown processes. Decomposition reactions are complicated and diverse, which makes them susceptible to the effect of several stimuli. These variables can be roughly divided into internal qualities of the chemical, such as its molecular structure, stability, and reactivity, and external circumstances, such as temperature, pressure, and the presence of catalysts.

The rate of decomposition and the amount of energy needed for the reaction are significantly influenced by the environmental circumstances. For instance, temperature has an impact on the kinetic energy of molecules, with greater temperature often resulting in more frequent collisions and enhanced molecular motion. As a result, decomposition reactions are frequently accelerated by greater temperature. On the other hand, pressure, especially in gaseous systems, may influence decomposition by changing the equilibrium between reactants and products (Lucht, 2017).

Catalysts or chemicals that speed up a chemical reaction without being consumed, can also have a significant impact on how decomposition reactions proceed. By offering a different reaction pathway with a lower activation energy, catalysts improve the breakdown process. They can speed up the breakdown process and allow the reaction to take place in more accommodating circumstances.

Internally, a compound's molecular make-up and structure have an impact on how susceptible it is to decomposition. Due to the existence of strained or weak chemical connections, some compounds have an intrinsic instability that makes them more prone to disintegrating. Others might have certain functional groups or substituents that boost reactivity or facilitate particular pathways for breakdown. In addition, the decomposition process may be impacted by the presence of pollutants, impurities or other substances. These compounds may function as catalysts, promote or hinder particular reactions, or change the reaction's overall mechanism (Chen, 2010).

It is important to remember that the variables affecting decomposition responses are interconnected and may interact. For instance, temperature impacts a compound's stability in addition to its effect on kinetic energy. A catalyst can reduce the impact of pressure or temperature, allowing breakdown to occur under more tolerable circumstances. Scientists can create and optimize reaction circumstances, forecast compound behavior, and create strategies for regulating and influencing decomposition processes by being aware of the variables that affect decomposition reactions. Numerous scientific fields, including chemistry, materials science, environmental science, and industrial processes, find uses for this knowledge.

The study of decomposition reactions necessitates a thorough comprehension of the variables that affect the rate, scope, and circumstances of these events. The molecular structure, stability, and reactivity of the chemicals as well as external factors like temperature, pressure, and catalysts work together to influence how breakdown processes behave. Scientists can obtain insights into the underlying mechanisms and create plans to regulate and optimize the decomposition processes for theoretical and practical applications by unraveling these elements.

5.3.1. Temperature

The rate and extent of decomposition processes are significantly influenced by temperature, which is a crucial element. Scientists can manage the energy supplied to the interacting molecules and modify the likelihood of successful collisions by changing the temperature, which affects the breakdown process. The kinetic energy of molecules increases with temperature, leading to quicker molecular motion. The reacting species collide more frequently and energetically as a result of this increased velocity. This implies that at higher temperatures, chemical bonds are broken and new bonds are formed more quickly in decomposition reactions (Chen, 2010).

The relationship between temperature and the rate of a decomposition process is described by the Arrhenius equation, which claims that the rate constant (k) of a reaction exponentially increases with temperature. This implies that even a small increase in temperature can have a large impact on the rate of breakdown. In addition to change the rate of disintegration, temperature can change how stable the molecules are. Some materials are more stable at lower temperatures, which makes them less prone to degradation. However, as the temperature rises, the stability of these compounds may decrease, making them more prone to breakdown (Lucht, 2017).

On the other side, some substances could be naturally unstable, and higher temperatures are more favorable for their disintegration. This may be because the molecule contains weak or strained bonds, which are more likely to dissolve at high temperatures. Additionally, a decomposition reaction's activation energy is influenced by temperature. The least amount of energy necessary for a reaction to take place is known as activation energy. Higher temperatures can break through the activation energy barrier by supplying enough thermal energy, speeding up the breakdown process.

It is very important to remember that decomposition reactions can also be negatively impacted by excessive temperatures. Unwanted side reactions or even total combustion, where the molecule combines with oxygen from the air instead of going through the intended breakdown, might result from excessively high temperatures. Similar to how high temperatures can make molecules less mobile and reactive, slowing down the breakdown process. For planning and enhancing reaction conditions in practical applications, it is essential to comprehend how decomposition reactions respond to temperature changes. By adjusting the temperature, scientists may control reaction rates, forecast compound behavior, and guarantee the intended outcome of the breakdown process.

In conclusion, a key element that strongly affects the decomposition reactions is temperature. It has an impact on the reaction's rate, the compounds' stability, and the amount of activation energy needed for decomposition. Scientists can regulate the energy available to the interacting molecules and the likelihood of successful collisions by adjusting the temperature, eventually regulating the course of decomposition reactions (Chen, 2010).

5.3.2. Catalysts

Because they offer a different reaction pathway with a lower activation energy, catalysts have a substantial impact on how decomposition reactions behave. They are extremely helpful in regulating and optimizing chemical processes since they speed up the reaction without being consumed themselves.

Catalysts in decomposition processes make it easier for a chemical to break down by lowering the energy barrier needed for the reaction to happen. They accomplish this by interacting with the molecules of the reactants, stabilizing transition states, or changing the reaction's process (Lucht, 2017).

By offering a different reaction pathway, catalysts can affect decomposition reactions in several ways. The reactant molecules interact with the catalyst, enabling an alternative set of intermediate steps to be used for their breakdown. This alternate route frequently entails the creation of transient bonds between the catalyst and the reactant, which helps the compound's chemical bonds to break. Catalysts reduce the activation energy needed for the decomposition reaction by offering this alternate pathway. The energy barrier that reactant molecules must cross to start the reaction is represented by activation energy. Catalysts make it possible for the breakdown to happen more quickly, at lower temperatures, or with less energy input. This occurs by lowering the activation energy.

Through the stabilization of transition states, catalysts can also affect decomposition reactions. High-energy intermediate entities called transition states are present during chemical reactions. These transition states can interact with catalysts, changing their electrical and structural characteristics and stabilizing them. By stabilizing the system, less energy is needed to enter the transition state, which speeds up the decomposition procedure. Catalysts can also alter the reaction's actual mechanism. In order to promote selective bond-breaking and facilitate the intended breakdown routes, they can offer reactive sites or functional groups that interact with particular regions of the reactant molecule. Catalysts can improve the selectivity, efficiency, and control of decomposition reactions by affecting the reaction mechanism.

It's vital to remember that catalysts don't change the decomposition reaction's overall thermodynamics. The finished goods or the equilibrium state are unaffected. Instead, catalysts quicken the reaction's pace by offering an energetically advantageous pathway that enables it to achieve the same equilibrium state more quickly. The specific breakdown reaction and the desired result will choose the catalyst to use. For different chemicals and reaction settings, various catalysts display differing degrees of efficacy and selectivity. Researchers frequently examine several catalysts and their properties to find the best catalyst for a specific decomposition reaction. (Chen, 2010).

There are various benefits of using catalysts in decomposition reactions. It can speed up reactions, use less energy, increase yield, and allow decomposition to take place under more comfortable settings. Additionally, catalysts offer ways to regulate reaction selectivity and reduce the production of undesirable byproducts. In conclusion, catalysts have a significant role in how decomposition processes develops. Catalysts quicken the rate of decay without being consumed themselves by offering substitute reaction pathways with lower activation energies. To support the chosen breakdown pathways, they interact with reactant molecules, stabilize transition states, and alter reaction processes. Scientists can better control chemical transformations and create and optimize processes by understanding the role of catalysts in decomposition reactions (Lucht, 2017).

5.3.3. Pressure

Particularly in gaseous systems, pressure can affect the decomposition reactions. Even though pressure has a less substantial effect on decomposition than temperature and catalysts do, it nevertheless affects the rate and size of the reaction. The equilibrium between the reactants and products in gaseous decomposition reactions can be impacted by pressure. Le Chatelier's principle states that when a system is in equilibrium and a change in pressure occurs, the equilibrium adjusts to account for the change and create a new equilibrium. This idea is applicable to gas-based degradation reactions.

The system makes an effort to reduce all of the gas molecules when the pressure is raised in an effort to lessen the pressure change. By choosing the breakdown reaction that produces the fewest moles of gas, this can be accomplished. On the other hand, if the pressure drops, the system tends to favor the decomposition reaction, which produces more gas molecules and raises the pressure. It is crucial to remember that pressure effects are typically more obvious at high pressures or in systems where the gas volumes of the reactants and products differ noticeably. The pressure dependency plays a relatively minor role in many decomposition reactions when compared to other elements.

Pressure has an impact on the equilibrium as well as the frequency of collisions and molecular interactions inside a system. An increased frequency of collisions results from the closer packing of the molecules as pressure rises. The likelihood of successful collisions may increase due to the increasing collision frequency, which could boost the decomposition rate. Additionally, the solubility of the gases and reactants involved in the decomposition reaction can be impacted by pressure. Changes in solubility can affect the concentration and availability of reactants, which can affect the total reaction rate. It is important to note that systems with gaseous reactants and products tend to exhibit pressure effects on decomposition processes more frequently. Pressure is frequently insignificant or extremely important in solid-state or condensed-phase reactions (Chen, 2010).

In general, pressure have an impact on decomposition processes even though it is not frequently the main determining factor. In gaseous systems, where pressure variations can affect equilibrium, collision frequency, and reactant solubility, the pressure-dependent effects are more prominent. However, in order to completely comprehend and regulate decomposition reactions, pressure must be taken into account in addition to other key variables like temperature and catalysts.

5.3.4. Light

Photolysis, also known as photolytic breakdown, is a process through which light can affect the decomposition events. When light energy is absorbed by a substance, photolysis happens. This causes chemical bonds to break and new species to form. Certain substances can be made active and breakdown reactions can start when light is present, which wouldn't happen otherwise. A compound's ability to absorb light depends on two factors: its electronic structure and the existence of chromophores, functional groups or conjugated systems that can absorb particular light wavelengths. A compound may experience structural alterations, bond breakages, or bond rearrangements as it absorbs light energy, leading to disintegration (Lucht, 2017).

When it comes to certain compounds or groups of compounds, the influence of light on breakdown reactions can be very specialized. While some substances may be very sensitive to particular light wavelengths, others may need stronger light sources or intensities to begin breaking down. Studies of environmental deterioration as well as photochemistry, photovoltaics, and other sectors all use light-induced breakdown reactions.

5.3.5. Concentration

The rate and scope of decomposition processes are greatly influenced by concentration, specifically the concentration of reactants. One of the key variables controlling reaction speeds is the frequency of molecular collisions, which is influenced by the concentration of reactants. For a chemical reaction to take place, reactant molecules must collide with enough force and in the right direction with each other. Higher reactant concentrations improve the likelihood of successful collisions since there are more molecules available for collision. In general, the reaction rate of breakdown increases as reactant concentration increases. This is due to the fact that larger concentrations result in more frequent bond breaks and breakdown events since they increase collision frequency and the likelihood of effective collisions (Chen, 2010).

It is crucial to remember that the effect of reactant concentration on reaction rate has a maximum. When reactant molecules are present in extremely high concentrations, they cluster together, increasing the likelihood of collisions, intermolecular interactions, and inhibitions. The assumed linear relationship between concentration and reaction rate may not hold true as a result of these factors. Additionally, the concentration of other species, such as inhibitors or catalysts, may have an effect on how quickly decomposition reactions occur. While inhibitors may slow the rate of decomposition by interfering with the reaction mechanism or stabilizing the reactant molecules, catalysts speed up the process by providing alternative reaction paths or lowering the activation energy (Lucht, 2017).

5.4. PRACTICAL APPLICATIONS OF DECOMPOSITION REACTIONS

Decomposition reactions have many real-world applications in a variety of businesses and scientific disciplines. It is possible to make improvements in the synthesis of new materials, the production of energy, the treatment of waste, and environmental remediation by being able to manage and use the breakdown of complex compounds into simpler ones. The following are some real-world uses for decomposition reactions:

Decomposition reactions are important in the synthesis of many different materials. For instance, thermal breakdown of precursor chemicals in the ceramics industry can be employed to produce ceramic powders with specified qualities. Similar to how metal oxides, nanoparticles, and thin films for electrical and optical devices are created, decomposition reactions are also used in these processes. Scientists can modify the structure, content, and characteristics of materials for particular uses by carefully regulating the breakdown conditions.

The production of energy is significantly impacted by decomposition reactions. For instance, fossil fuels like coal or natural gas are thermally broken down in power plants to produce electricity. Thermochemical reactions for solar energy storage and conversion, as well as the steam reforming of hydrocarbons to produce hydrogen, both depend on decomposition reactions.

These techniques aid in the production of sustainable energy by capturing the energy released during decomposition (Chen, 2010).

Explosives and pyrotechnics both use decomposition reactions as key components of their mechanisms. These reactions include the carefully controlled breakdown of particular molecules to release gases, light, heat, and sound. Pyrotechnic displays, fireworks, and explosive materials can be created for amusement, military uses, mining, and building by choosing the right reactants and comprehending their breakdown properties (Lucht, 2017).

1. **EnvironmentalRemediation:** Wastetreatmentandenvironmental remediation both include decomposition reactions. For instance, pyrolysis and incinerator processes use heat breakdown to manage hazardous or organic waste. Pollutants and contaminants can be transformed into less dangerous or inert chemicals through decomposition processes. Additionally, organic contaminants are removed from water by means of chemical decomposition processes, and complex substances are broken down into more basic, non-toxic forms.
2. **Industries of the Pharmaceutical and Chemical Kind:** Decomposition processes have real-world uses in these fields. They utilized in the synthesis of pharmaceutical medications, in which certain reactions are exploited to transform starting materials into desired active components. Decomposition processes are also used to make intermediates, polymers, and specialized compounds for a variety of industrial uses (Chen, 2010).

In conclusion, decomposition reactions are fascinating chemical processes that aid in both our understanding of the macro- and micro-scales of the Earth. These interactions have a wide range of practical ramifications, from industrial uses to environmental and ecological situations. We can understand the basics, contributing factors, types, and applications of decomposition reactions, as well as their importance in a variety of scientific fields and daily life. Researchers and practitioners can use the power of decomposition reactions to propel advancement and innovation in a variety of disciplines by exercising caution and adhering to safety precautions (Lucht, 2017).

CHAPTER

6 Combustion Reactions

Contents

6.1. INTRODUCTION

Combustion reactions, also referred to as combustion processes, are fundamental chemical reactions characterized by their exothermic nature. They involve the rapid combination of a fuel and an oxidizer, leading to the liberation of energy in the form of heat and light. These reactions commonly occur when carbon-based compounds, particularly hydrocarbons, undergo oxidation by interacting with oxygen, which acts as the primary oxidizer.

Combustion reactions play a vital role in various natural and human-made processes. They are essential for numerous energy conversion systems, such as combustion engines, power plants, and even the simple act of burning wood for warmth. Understanding combustion reactions is crucial for advancing our knowledge of energy production, environmental impacts, and fire safety (Pfefferle, 2012).

The fuel component of a combustion reaction is mostly a hydrocarbon compound, which consists of hydrogen and carbon atoms bonded together. Hydrocarbons are abundant in nature and can be found in fossil fuels like gasoline, natural gas, coal, and oil. These compounds possess high chemical potential energy, which can be released through combustion.

Oxygen acts as the primary oxidizer in combustion reactions. It combines with the fuel, breaking the chemical bonds within the hydrocarbon molecules. The breaking of these bonds releases a significant amount of energy stored in the fuel. The combustion process converts the chemical potential energy of the fuel into thermal energy, manifesting as heat, and radiant energy in the form of light (Mukasyan, 2014).

The oxidation of hydrocarbons in combustion reactions involves a highly exothermic reaction. As the fuel and oxidizer combine, the reaction becomes self-sustaining, releasing energy that accelerates the reaction rate. This self-sustaining nature is why combustion reactions often exhibit a characteristic flame or glowing effect.

The products of combustion reactions depend on the specific fuel being burned. In general, hydrocarbon combustion reactions result in the formation of CO_2 and water (H_2O) as the primary products. Other combustion byproducts, such as carbon monoxide (CO), nitrogen oxides (NOx), sulfur dioxide (SO_2), and particulate matter, can also be formed depending on the composition of the fuel and the combustion conditions (Pfefferle, 2012) (Figure 6.1).

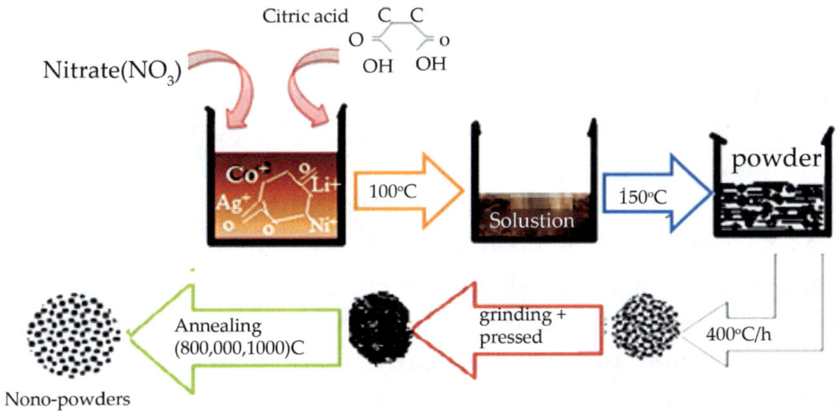

Figure 6.1. *Schematic of combustion reaction.*

Source: Haider, A., Al-Tabbakh, A., Al-Zubaidi, A., & Rsool, R., (2018). Preparation and characterization of $LiCo_{0.5}Ni_{0.45}Ag_{0.05}O_2$ cathode material for lithium–ion battery. Journal of Materials Science: Materials in Electronics. – Scientific Figure on ResearchGate. Available from: https://www.researchgate.net/figure/Schematic-of-combustion-reaction_fig1_325700705.

Combustion reactions have a significant impact on the environment. The release of CO_2 during the combustion of fossil fuels contributes to the greenhouse effect and climate change. The formation of other byproducts, such as nitrogen oxides and particulate matter, can contribute to air pollution and have adverse effects on human health and the environment. Therefore, understanding and optimizing combustion processes are essential for minimizing environmental impacts and developing cleaner and more efficient energy systems.

6.1.1. Importance of Combustion Reactions

Combustion reactions play a vital role in numerous aspects of our lives, making them of utmost importance. These reactions involve the quick combination of a fuel source with oxygen, resulting in the release of energy in the form of heat and light. This process has been harnessed by humanity for various applications, ranging from providing heat and electricity to powering vehicles and industrial processes. The significance of combustion reactions can be observed in multiple areas, including energy production, transportation, and even everyday activities (Mukasyan, 2014). One of the primary reasons combustion reactions are essential is their role in energy production. Fossil fuels, such as coal, oil, and natural gas, serve as the primary sources of energy for electricity generation in many parts of the world. When these fuels undergo combustion, the chemical bonds holding their carbon and hydrogen atoms break, allowing them to combine with oxygen. This chemical reaction releases a substantial amount of heat energy, which can be converted into electrical energy through various mechanisms, such as steam turbines or internal combustion engines.

Transportation heavily relies on combustion reactions as well. The majority of vehicles, including cars, trucks, ships, and airplanes, use internal combustion engines that burn gasoline or diesel fuel to propel them forward. The combustion of these fuels within the engines generates high temperatures and rapidly expanding gases, which in turn produce the necessary force to move the vehicle. Without combustion reactions, our modern transportation systems would not exist in their current form, leading to significant limitations in terms of mobility and logistics (Pfefferle, 2012).

Moreover, combustion reactions are crucial for everyday activities and household applications. Many households depend on combustion-based systems for heating and cooking purposes. Whether it's a gas stove in the kitchen or a furnace that provides warmth during the months of winter, these systems rely on controlled combustion reactions to release heat energy. Additionally, combustion-based technologies are utilized in water heaters, boilers, and even fireplaces exemplifying the importance of combustion reactions in our daily lives.

Beyond energy production and transportation, combustion reactions have significant implications for various industries. For instance, combustion is an essential process in manufacturing cement, iron, steel, and other building materials. These industries rely on controlled combustion reactions to achieve high temperatures required for chemical transformations and material synthesis. Additionally, combustion reactions are employed in waste management systems, where controlled burning of waste materials helps reduce their volume and convert them into less harmful forms (Mukasyan, 2014).

However, while combustion reactions provide valuable benefits, it is crucial to acknowledge their environmental impact. The burning of fossil fuels release greenhouse gases, such as CO_2, contributing to climate change and air pollution. Therefore, it becomes imperative to explore alternative fuel sources and cleaner combustion technologies to mitigate these environmental concerns.

Combustion reactions hold immense importance in various aspects of our lives. They are fundamental for energy production, transportation, household applications, and numerous industries. Recognizing the significance of combustion reactions encourages us to explore sustainable alternatives and adopt cleaner technologies to minimize their environmental impact.

6.2. PRINCIPLES OF COMBUSTION REACTIONS

Combustion reactions operate based on a set of principles that govern their behavior and allow for efficient energy release. These principles can be understood by examining the components involved and the role of energy in the process.

6.2.1. Fuel and Oxidizer

The two key components necessary for a combustion reaction are the fuel and the oxidizer. The fuel serves as an essential energy source, providing the necessary chemical compounds that can undergo combustion. Common examples of fuels include hydrocarbons like gasoline, natural gas, coal, and biomass. These fuels contain high-energy carbon and hydrogen atoms that can be oxidized.

On the other hand, the oxidizer, most commonly oxygen from the surrounding air, supplies the oxygen atoms required for the reaction to occur. Oxygen is an excellent oxidizer due to its high electronegativity and ability to readily accept electrons from other elements. The oxidizer facilitates the breaking of the chemical bonds within the fuel molecules, allowing them to combine with oxygen and release energy in the form of heat and light (Pfefferle, 2012).

Understanding the characteristics of different fuels and oxidizers is essential for optimizing combustion processes. Different fuels have varying energy contents and combustion characteristics, which can affect the efficiency and performance of combustion systems. Similarly, the availability and purity of the oxidizer can influence the combustion process and the resulting emissions.

6.2.2. The Role of Energy and Activation Energy

Energy is a fundamental factor in combustion reactions. To initiate a combustion reaction, an initial input of energy is required. This energy is known as activation energy and is necessary to break the existing chemical bonds within the fuel molecules. Once the activation energy threshold is reached, the reaction proceeds spontaneously, releasing a significant amount of energy (Mukasyan, 2014).

The activation energy can be thought of as a barrier that must overcome for the reaction to occur. It is influenced by factors such as temperature, pressure, and the chemical nature of the fuel and oxidizer. Higher temperatures generally result in lower activation energies, making the reaction more favorable and efficient.

Understanding energy transfer mechanisms within combustion systems is crucial for optimizing their performance. Efficient energy transfer ensures that a larger proportion of the released energy is harnessed for useful purposes, such as generating electricity or powering mechanical systems. By minimizing energy losses, combustion processes can be made more efficient and environmentally friendly (Figure 6.2).

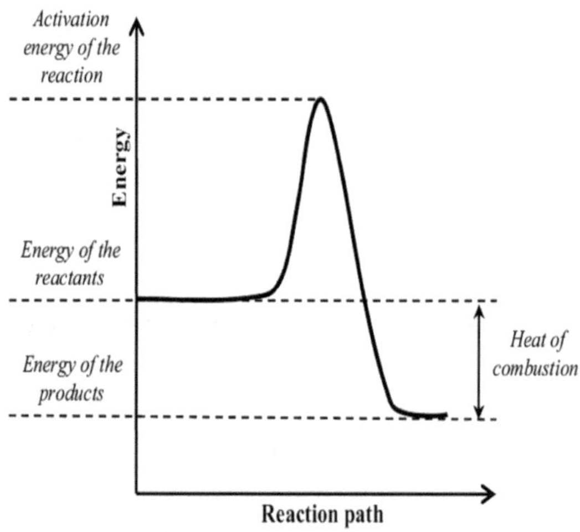

Figure 6.2. *Activation energy of a reaction.*

Source: Elbasuney, S., Gobara, M., & Yehia, M., (2019). Ferrite nanoparticles: Synthesis, characterization, and catalytic activity evaluation for solid rocket propulsion systems. Journal of Inorganic and Organometallic Polymers and Materials. – Scientific Figure on ResearchGate. Available from: https://www.researchgate.net/figure/Schematic-for-activation-energy-of-high-energetic-systems-sufficient-energy-activation_fig1_329627474.

Furthermore, knowledge of the factors affecting activation energy allows engineers and scientists to design combustion systems with lower energy requirements. By reducing the activation energy, the efficiency and effectiveness of combustion reactions can be enhanced, resulting in improved energy utilization and reduced waste (Pfefferle, 2012).

6.2.3. The Combustion Triangle

Understanding the principles of combustion reactions begins with recognizing the fundamental concept known as the "combustion triangle." The combustion triangle represents the three essential elements required for combustion to occur: fuel, oxygen, and an ignition source. These elements must be present in specific proportions for combustion to initiate and sustain.

The first element of the combustion triangle is fuel. Fuel refers to any substance that can undergo combustion and release energy. Common examples of fuels include hydrocarbons like gasoline, diesel, natural gas, and coal. However, other materials, such as wood, biomass, and even certain metals, can also serve as fuels under specific conditions. The energy stored in the chemical bonds of the fuel is released when the combustion reaction takes place (Mukasyan, 2014).

The second element of the combustion triangle is oxygen. Oxygen is a critical component necessary for the combustion process. It serves as the oxidizer, which means it combines chemically with the fuel during combustion. The presence of oxygen enables the fuel to undergo oxidation, resulting in the release of heat energy and the formation of combustion byproducts, such as CO_2, water vapor, and various pollutants.

The third element of the combustion triangle is an ignition source. An ignition source provides the initial energy required to start the combustion reaction. It can take various forms, such as an open flame, a spark, a hot surface, or even a high temperature. The ignition source must provide enough amount of energy to overcome the activation energy barrier of the combustion reaction, allowing the fuel and oxygen to react and initiate the self-sustaining process of combustion (Figure 6.3).

Figure 6.3. *Combustion triangle.*

Source: Gorbett, G. E., & Kozhumal, S. P., (2023). Fire fundamentals. In: Meacham, B. J., & McNamee, M., (eds.), Handbook of Fire and the Environment. The society of fire protection engineers series. Springer, Cham. https://doi.org/10.1007/978-3-030-94356-1_3.

The combustion triangle illustrates the interdependence of these three elements. If any one of the elements is absent or insufficient, combustion cannot occur or will be limited. For example, in the absence of oxygen, known as oxygen deprivation, a fire cannot be sustained even if fuel and an ignition source are present. Similarly, if the fuel source is exhausted or the ignition source is removed, the combustion process will cease (Pfefferle, 2012).

6.2.4. Types of Combustion Reactions

Combustion reactions can be categorized into different types based on the characteristics of the fuel, the conditions under which the combustion occurs, and the resulting products. Here are a few notable types of combustion reactions:

1. **Complete Combustion:** In complete combustion, the fuel combines with oxygen in ample supply, resulting in the production of CO_2 and water vapor (H_2O) as the primary combustion products. This type of combustion is considered efficient since it maximizes the release of energy from the fuel.

2. **Incomplete Combustion:** In contrast to complete combustion, incomplete combustion occurs when the fuel does not have access to an adequate oxygen supply. This results in the production of carbon monoxide (CO) and soot (partially burnt carbon) as byproducts. Incomplete combustion is often less efficient and can lead to the release of harmful pollutants into the environment (Mukasyan, 2014).
3. **Rapid Combustion:** Also known as explosive combustion, involves a rapid release of energy and the production of a shockwave. It occurs when the combustion process happens extremely quickly, leading to an explosive force. Explosives and fireworks are examples of substances that undergo rapid combustion.
4. **Spontaneous Combustion:** It occurs when a substance ignites and burns without an external ignition source. This phenomenon typically happens when certain materials, such as oily rags, hay, or coal dust, undergo slow oxidation and reach a critical temperature, leading to self-ignition.
5. **Combustion of Metals:** Some metals, such as magnesium and aluminum, can undergo combustion reactions under specific conditions. When ignited, these metals react with oxygen to form metal oxides and release a significant amount of heat energy. Combustion of metals is often used in pyrotechnics and welding processes.

Understanding the different types of combustion reactions allows us to better comprehend the behavior of different fuels, the conditions required for combustion, and the potential environmental impacts associated with each type. By applying this knowledge, scientists and engineers can develop more efficient combustion systems and strategies, as well as implement measures to mitigate the adverse effects of combustion, such as reducing pollutant emissions and promoting cleaner energy sources.

6.3. COMBUSTION MECHANISMS

6.3.1. Initiation Stage

The initiation stage of combustion marks the beginning of a fascinating process where energy is applied to overcome the activation energy barrier, igniting a sequence of events that lead to combustion. This crucial stage acts as the catalyst, triggering the reaction and initiating the formation of highly reactive species known as radicals (Pfefferle, 2012).

During the initiation stage, an external energy source is introduced to the system, such as heat, a spark, or a flame. This energy is essential in providing the necessary activation energy to break the bonds within the reactant molecules. Once this energy threshold is surpassed, the reactant molecules become destabilized, resulting in the formation of radicals.

Radicals are highly reactive species characterized by the presence of unpaired electrons. These unpaired electrons make radicals extremely reactive, seeking out stable electron configurations by reacting with other molecules in their vicinity. The formation of radicals during the initiation stage is crucial as they are driving force behind the subsequent stages of combustion (Mukasyan, 2014).

6.3.2. Propagation Stage

Following the initiation stage, the combustion process enters the propagation stage. This stage is characterized by a chain reaction involving the continuous reaction between radicals, fuel molecules, and oxidizer molecules. The propagation stage is of utmost importance as it sustains combustion, allowing for the release of energy and the production of additional radicals.

During propagation, radicals generated in the initiation stage react with fuel molecules, such as hydrocarbons, and oxidizer molecules, typically oxygen from the air. These reactions result in the formation of intermediate species and more radicals. The newly formed radicals then proceed to react with more fuel and oxidizer molecules, perpetuating the chain reaction.

The chain reaction in the propagation stage plays a pivotal role in releasing energy. As the reaction progresses, the energy released by the formation of new bonds outweighs the energy required to break the initial bonds. This excess energy is released in the form of heat and light, contributing to the characteristic flames and heat associated with combustion processes.

Furthermore, the production of additional radicals in the propagation stage ensures that the chain reaction continues, sustaining the release of energy. This continuous propagation of radicals allows for the efficient consumption of fuel and ensures that combustion proceeds in a self-sustaining manner.

6.3.3. Termination Stage

As the combustion process evolves, the termination stage eventually follows the propagation stage. In this stage, radical concentrations start to decrease, and the combustion reaction gradually slows down or ceases. Understanding the termination stage is crucial for optimizing combustion efficiency and preventing undesired side reactions (Pfefferle, 2012).

In the termination stage, various factors contribute to the decline in radical concentrations. One of the primary reasons is the depletion of reactant molecules. As the fuel and oxidizer are consumed during the propagation stage, the availability of these molecules decreases, resulting in fewer opportunities for radical formation.

Additionally, the termination stage is influenced by various factors such as temperature, pressure, and the presence of inhibitors or quenchers. Lower temperatures and higher pressures tend to favor radical recombination, leading to a more rapid decline in radical concentrations. Inhibitors or quenchers are substances that can react with radicals, effectively terminating the chain reaction and slowing down the combustion process.

Understanding the termination stage is essential for optimizing combustion processes, as it allows for the control and prevention of unwanted side reactions. By carefully managing factors such as reactant concentrations, temperature, pressure, and the use of inhibitors, engineers and scientists improve combustion efficiency, reduce pollutant emissions, and ensure safe and effective combustion in various applications, ranging from engines to industrial processes (Mukasyan, 2014).

Overall, the combustion process encompasses the initiation, propagation, and termination stages. The initiation stage provides the necessary energy to overcome the activation barrier, resulting in the formation of highly reactive radicals. These radicals then drive the propagation stage, where they react with fuel and oxidizer molecules, sustaining the combustion process and releasing energy. Finally, the termination stage marks the decline in radical concentrations and the gradual slowing down of the combustion reaction. Understanding and controlling each of these stages is crucial for optimizing combustion efficiency, ensuring safety, and minimizing environmental impacts.

6.3.4. Reaction Kinetics in Combustion

Reaction kinetics plays a crucial role in unraveling the intricacies of combustion processes by shedding light on the speed at which chemical reactions occur. By studying the kinetics of combustion reactions, scientists and engineers gain valuable insights into the fundamental mechanisms governing the transformation of reactants into products during combustion. A deeper understanding of these mechanisms allows for the development of highly efficient combustion systems, leading to improved energy utilization and reduced environmental impacts.

Temperature stands out as a pivotal factor influencing the kinetics of combustion reactions. As the temperature increases, the average kinetic energy of the reactant molecules escalates, resulting in more frequent and energetic collisions. This enhanced collision frequency leads to a greater number of successful reactions, thereby accelerating the overall combustion process. Furthermore, elevated temperatures can facilitate the breaking of chemical bonds within reactant molecules, enabling the formation of new bonds and the generation of desired products. However, excessively high temperatures may also promote the formation of unwanted byproducts or cause thermal degradation of the reactants, necessitating careful temperature control for optimal combustion efficiency.

Pressure is another key parameter affecting the kinetics of combustion reactions. An increase in pressure generally leads to a higher collision frequency among gaseous reactant molecules, intensifying the chances of effective collisions and promoting reaction rates. Additionally, elevated pressure can modify the equilibrium composition of combustion systems, favoring certain reactions over others. Therefore, understanding the influence of pressure on reaction kinetics is vital for designing combustion systems with optimal operating conditions and achieving desired performance outcomes (Pfefferle, 2012).

Reactant concentrations play a vital role in determining the reaction kinetics of combustion processes. Higher reactant concentrations typically lead to increased collision frequencies and, consequently, enhanced reaction rates. Moreover, reactant concentration gradients within combustion systems can influence reaction pathways and selectivity, resulting in variations in the formation of different products. By carefully manipulating the reactant concentrations, combustion systems can be tailored to favor specific reactions, enabling greater control over the combustion process and optimizing energy conversion efficiencies (Mukasyan, 2014).

The presence of catalysts in combustion reactions significantly impacts on their kinetics. Catalysts facilitate reactions by lowering the activation energy required for reactant molecules to overcome the energy barrier and transform into products. This reduction in activation energy enables reactions to occur more readily, thereby accelerating the overall combustion process. Catalysts can also promote selectivity, guiding the reactions towards desired products while suppressing the formation of unwanted byproducts. Consequently, the strategic utilization of catalysts in combustion systems can enhance reaction kinetics, improve combustion efficiency, and mitigate environmental emissions.

Overall, the study of reaction kinetics in combustion is indispensable for comprehending the rate at which chemical transformations occur during the combustion process. By considering factors such as temperature, pressure, reactant concentrations, and catalysts, scientists and engineers can design combustion systems with enhanced efficiency, reduced environmental impact, and tailored product selectivity. Continued research in reaction kinetics promises to unlock further advancements in combustion science, leading to the development of cleaner and more sustainable energy technologies.

6.4. APPLICATIONS OF COMBUSTION REACTIONS

6.4.1. Internal Combustion Engines

Internal combustion engines are a vital innovation that has transformed the way we power various modes of transportation and even generate electricity. These engines are widely used in automobiles, motorcycles, airplanes, ships, and power generators due to their efficiency and versatility. By harnessing the power of combustion reactions, internal combustion engines play a crucial role in our modern world.

The principle behind internal combustion engines lies in controlled explosions occurring within a confined space. Fuel, typically gasoline or diesel, is mixed with air in the engine's combustion chamber. This mixture is then ignited by a spark plug, resulting in a rapid combustion reaction. The energy released during combustion forces the piston within the engine to move, converting the chemical energy into mechanical work. This motion ultimately propels the vehicle or drives a generator, producing electricity.

One of the primary applications of internal combustion engines is in transportation. The automobile industry heavily relies on these engines to power cars and trucks, providing efficient and reliable means of transportation. From everyday commuting to long-distance journeys, internal combustion engines have become an integral part of our mobility infrastructure. Their versatility allows them to be adapted for various vehicle sizes, power requirements, and fuel types, making them suitable for different transportation needs worldwide.

Moreover, internal combustion engines are widely utilized in the aviation industry. Airplanes, both commercial and private, rely on these engines to generate the necessary thrust for takeoff, cruising, and landing. The power produced by the combustion reactions enables aircraft to overcome gravity and remain airborne. Additionally, internal combustion engines play a crucial role in smaller aircraft, such as helicopters and drones, enabling vertical takeoff and precise maneuverability (Mukasyan, 2014).

Maritime transportation also heavily relies on internal combustion engines. Ships, ranging from small recreational boats to massive cargo vessels, utilize these engines for propulsion. Whether it's cruising through calm waters or navigating rough seas, internal combustion engines provide the necessary power to propel ships, making global trade and transportation efficient and accessible (Pfefferle, 2012).

Beyond transportation, internal combustion engines find application in power generation also. They are commonly used in generators to produce electricity for various purposes. These engines can be fueled by natural gas, diesel, or other liquid fuels, allowing for the reliable generation of electricity in both stationary and mobile setups. Internal combustion engine-powered generators are often employed in remote areas or during emergencies, ensuring a stable power supply when grid electricity is unavailable or disrupted.

While internal combustion engines have played a crucial role in our lives, there is increasing awareness about their environmental impact. The combustion of fossil fuels releases greenhouse gases, contributing to climate change. To address this concern, efforts are being made to develop cleaner and more efficient engines. Hybrid vehicles, electric cars, and alternative fuel sources such as hydrogen and biofuels are being explored to reduce emissions and improve sustainability.

6.4.2. Combustion in Power Generation

Combustion reactions play a pivotal role in power generation, serving as the primary source of energy in various industries and sectors worldwide. The application of combustion in power generation encompasses a wide range of technologies and fuels, each contributing to the generation of electricity, heat, or a combination of both.

One of the most prevalent methods of power generation through combustion is the utilization of fossil fuels, such as coal, natural gas, and petroleum. These fuels possess high energy density and readily combusted to produce heat. In traditional power plants, fossil fuels are burned in combustion chambers, producing high-pressure steam that drives turbines connected to electrical generators. The rotational motion of the turbines converts the thermal energy of combustion into mechanical energy, which is subsequently converted into electricity.

Modern power plants also employ advanced combustion technologies, such as fluidized bed combustion and integrated gasification combined cycle (IGCC). Fluidized bed combustion involves suspending fuel particles in a high-velocity stream of air or gas, creating a fluidized bed. This technology allows for efficient combustion with reduced emissions and enables the use of a wide range of fuels, including coal, biomass, and waste materials (Mukasyan, 2014).

IGCC, on the other hand, integrates gasification and combustion processes to generate power. In this method, the fuel is partially oxidized in a gasifier, producing a synthesis gas (syngas) consisting of carbon monoxide and hydrogen. The syngas is combusted in a gas turbine, and the resulting hot exhaust gases drive a steam turbine, generating electricity. IGCC offers higher efficiency compared to conventional combustion processes and can utilize a variety of fuels, including coal, biomass, and petroleum coke.

Furthermore, combustion is also employed in decentralized power generation systems, such as gas turbines and internal combustion engines. Gas turbines operate by compressing air and combusting fuel in a combustion chamber. The expanding hot gases drive the turbine, which is connected to a generator, producing electricity. These turbines can operate on various fuels, including natural gas, diesel, and biofuels (Pfefferle, 2012).

Internal combustion engines, commonly used in automobiles and small-scale power generation, rely on the controlled combustion of fuel-air mixtures. The combustion process drives the pistons, converting the chemical energy of the fuel into mechanical energy, which can be harnessed to power vehicles or small generators. These engines can run on a wide range of fuels, including gasoline, diesel, and compressed natural gas.

While fossil fuels have been the primary source of combustion-based power generation, there is a growing focus on utilizing renewable and alternative fuels to mitigate environmental impacts. Renewable sources such as biomass, biogas, and waste-to-energy technologies enable the sustainable combustion of organic materials, reducing greenhouse gas emissions and promoting circular economy.

6.4.3. Combustion for Heat and Light

Combustion reactions play a vital role in providing both heat and light, making them indispensable in various applications. From powering vehicles and generating electricity to cook food and illuminating our surroundings, combustion is a fundamental process that enables us to meet our daily energy needs.

One of the primary applications of combustion reactions is in the production of heat. Combustible substances, such as fossil fuels like coal, oil, and natural gas, as well as biomass and even certain metals, can undergo combustion to release energy in the form of heat. This heat can be harnessed for various purposes, including heating homes and buildings, industrial processes, and thermal power generation. In heating systems, combustion takes place within furnaces or boilers, where the fuel is burned to produce hot gases that are then used to raise the temperature of air or water, providing warmth for indoor environments or facilitating industrial processes (Mukasyan, 2014).

Another crucial application of combustion reactions is in the generation of light. Combustion-based lighting systems have been used for centuries to illuminate our surroundings. Traditional sources of light, such as candles, oil lamps, and gas lamps, rely on combustion to produce a flame that emits light. Although modern lighting technologies, such as incandescent bulbs, fluorescent lights, and light-emitting diodes (LEDs), have largely replaced an open flame combustion for illumination, the underlying principle of converting energy released through combustion into light remains the same.

In addition to heat and light, combustion reactions also have significant implications for transportation. Internal combustion engines, found in most automobiles, motorcycles, and various forms of transportation, rely on the controlled combustion of a fuel-air mixture to generate power. When the fuel is burned inside the engine's cylinders, the resulting expansion of gases pushes pistons, which in turn drives the vehicle forward. This application of combustion has revolutionized transportation, enabling us to travel quickly and efficiently over vast distances (Pfefferle, 2012).

Moreover, combustion reactions play a crucial role in power generation. In thermal power plants, fossil fuels such as coal, oil, or natural gases are burned to produce steam, which drives turbines connected to electricity generators. The combustion process releases energy that is ultimately converted into electrical energy, powering homes, businesses, and industries. While the environmental impact of such power generation methods has raised concerns, efforts are underway to explore cleaner and more sustainable alternatives, such as renewable energy sources like solar, wind, and hydroelectric power.

Combustion reactions also find applications in everyday activities like cooking. Whether it's a gas stove, a charcoal grill, or a wood-fired oven, the combustion of fuel is essential for heating and cooking food. The heat produced by the controlled combustion process allows us to boil, fry, bake, and roast, making it possible to prepare a wide range of delicious meals.

6.4.4. Industrial Applications of Combustion

Combustion reactions play a vital role in various industrial applications, contributing to the production of energy, manufacturing processes, and waste management. The exothermic nature of combustion, coupled with the abundance of combustible substances, makes it a versatile and widely utilized phenomenon in several sectors.

One of the primary applications of combustion is in the field of energy production. Combustion reactions, particularly those involving fossil fuels such as coal, oil, and natural gas, are extensively employed in power plants to generate electricity. These fuels are burned in combustion chambers, releasing heat energy that converts water into steam. These high-pressure steam drives turbines, which rotate generators, producing electricity. This process is commonly used in both traditional thermal power plants and advanced combined cycle power plants (Mukasyan, 2014).

Moreover, combustion reactions are employed in internal combustion engines, which power various modes of transportation such as cars, motorcycles, airplanes, and ships. In these engines, controlled combustion of a fuel-air mixture occurs within the cylinders, resulting in the expansion of gases that ultimately move the pistons. The rotational motion generated by the pistons is converted into mechanical energy, propelling the vehicles forward. The development of more efficient and environmental friendly combustion engines has been a significant area of research and innovation in recent years.

Beyond energy production, combustion reactions find extensive use in manufacturing processes. Industrial furnaces and kilns utilize combustion to achieve high temperatures necessary for processes like metal smelting, glass manufacturing, and ceramic production. By burning fuels or using other combustible materials, these furnaces provide the heat required for melting raw materials, shaping molten metals, and creating durable products. The precise control of combustion parameters, such as air-fuel ratios and temperature profiles, is crucial to ensure optimal product quality and process efficiency.

Another important application of combustion lies in waste management. Incineration, a controlled combustion process, is employed to dispose solid waste materials in a safe and environmentally responsible manner. Waste-to-energy plants utilize this process to convert municipal solid waste into heat and electricity. By combusting the waste at high temperatures, organic matter is broken down, and the energy released is harnessed for power generation. Advanced technologies and strict emission control measure ensure that harmful byproducts are minimized, making waste incineration a viable solution for reducing landfill space and harnessing energy from waste (Pfefferle, 2012).

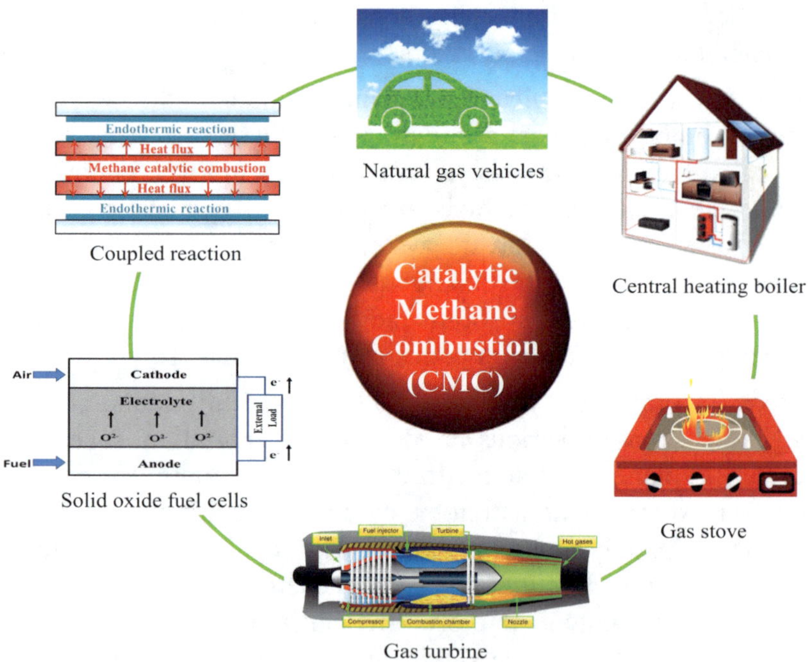

Figure 6.4. *Some of the application of combustion reactions.*

Source: He, L., Fan, Y., Bellettre, J., Yue, J., & Luo, L., (2019). A review on catalytic methane combustion at low temperatures: Catalysts, mechanisms, reaction conditions and reactor designs. Renewable and Sustainable Energy Reviews. – Scientific Figure on ResearchGate. Available from: https://www.researchgate.net/figure/Main-applications-of-CMC_fig3_337676710.

Additionally, combustion reactions have diversified applications in sectors such as agriculture, chemical production, and material science. In agricultural practices, controlled burning of crop residues is performed to clear fields, eliminate pests, and replenish soil nutrients. In chemical industries, combustion is employed in processes like steam cracking, which breaks down hydrocarbon molecules into smaller fragments for the production of various chemicals and plastics. Moreover, combustion synthesis is used in materials science to create ceramic powders and coatings with unique properties (Mukasyan, 2014) (Figure 6.4).

6.5. ENVIRONMENTAL IMPLICATIONS

6.5.1. Air Pollution and Greenhouse Gas Emissions

Combustion processes are a significant contributor to air pollution and greenhouse gas The combustion processes involve in various industries, transportation, and energy generation are major contributors to air pollution and greenhouse gas emissions. These emissions include CO_2, nitrogen oxides (NOx), sulfur dioxide (SO_2), and particulate matter (PM). The impact of these pollutants on the environment and human health cannot be overlooked. Therefore, understanding their sources, formation mechanisms, and control strategies becomes crucial in mitigating their adverse effects (Pfefferle, 2012).

When fossil fuels are burned, such as coal, oil, and natural gas, they release CO_2 into the atmosphere. CO_2 is a greenhouse gas that contributes to the trapping of heat within the Earth's atmosphere, leading to global warming and climate change. Additionally, combustion processes can release nitrogen oxides and sulfur dioxide, which contribute to the formation of smog, acid rain, and other harmful air pollutants.

Particulate matter, another byproduct of combustion, refers to tiny particles suspended in the air, often referred to as PM2.5 or PM10, depending on their size. These particles can have detrimental effects on human health when inhaled, leading to respiratory problems, cardiovascular diseases, and other adverse health outcomes (Mukasyan, 2014).

To address these environmental challenges, it is essential to develop effective strategies for reducing air pollution and greenhouse gas emissions.

This includes implementing stricter regulations and standards for industrial emissions, adopting cleaner technologies, and promoting the use of renewable energy sources. For example, transitioning from coal-fired power plants to cleaner alternatives like natural gas, solar, wind, or hydroelectric power can significantly reduce both air pollution and greenhouse gas emissions.

6.5.2. Combustion and Climate Change

The combustion of fossil fuels plays a major role in driving climate change. As mentioned earlier, the burning of these fuels releases greenhouse gases, primarily CO_2, into the atmosphere. These gases act as a blanket, trapping heat from the sun and preventing it from escaping back into space. This phenomenon is known as the greenhouse effect, leads to a gradual increase in global temperatures, causing climate patterns to change and resulting in a wide range of environmental and socioeconomic impacts.

The combustion of fossil fuels for electricity generation and transportation is the largest source of greenhouse gas emissions globally. Power plants that rely on fossil fuels, such as coal-fired plants, emit substantial amounts of CO_2. Similarly, vehicles powered by gasoline or diesel engines release CO_2 and other pollutants into the air. Taking an example of aircraft internal engine combustion, see the illustration below (Figure 6.5):

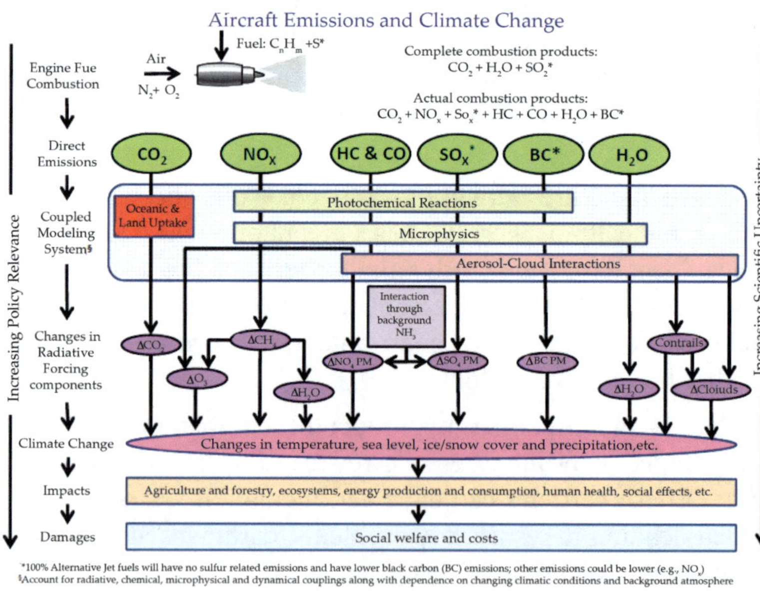

Figure 6.5. *Climate change as impacted by combustion emissions from aircraft.*

Source: Brasseur, et al., (2016). Impact of aviation on climate: FAA's aviation climate change research initiative (ACCRI), Phase II. Bulletin of the American Meteorological Society. – Scientific Figure on ResearchGate. Available from: https://www.researchgate.net/figure/Schematic-representation-of-emissions-from-aircraft-combustion-and-their-potential_fig1_312193345.

To mitigate the adverse effects of combustion-related climate change, there is an essential need to shift towards cleaner energy sources. Renewable energy technologies, such as solar and wind power, produce electricity without emitting greenhouse gases. Electrification of transportation by promoting electric vehicles and improving public transportation systems can also significantly reduce emissions associated with combustion engines (Pfefferle, 2012).

Furthermore, advancements in energy storage systems, smart grids, and energy efficiency measures can enhance the integration of renewable energy sources into the existing energy infrastructure. These measures not only help to reduce greenhouse gas emissions but also in promoting energy security, economic growth, and sustainable development.

6.5.3. Combustion Efficiency and Emissions Control

Improving combustion efficiency is a key strategy for minimizing fuel consumption and reducing emissions. When combustion processes are optimized, less fuel is required to generate the same amount of energy, leading to lower greenhouse gas emissions and reduced dependence on fossil fuels.

Several techniques and technologies are employed to enhance combustion efficiency and control emissions. Combustion optimization involves adjusting variables such as air-to-fuel ratios, burner design, and temperature control to maximize the efficiency of the combustion processes. This not only reduces fuel waste but also helps to minimize the release of pollutants.

Advanced combustion systems, such as fluidized bed combustion and combined cycle power plants, offer higher efficiency and lower emissions compared to conventional combustion technologies. These systems employ innovative designs and processes to extract more energy from fuel while reducing the release of pollutants.

Emissions control technologies play a crucial role in mitigating the environmental impact of combustion processes. These technologies include gas desulfurization systems, selective catalytic reduction, electrostatic precipitators, and particulate filters. They work to remove or transform pollutants before they released into the atmosphere, helping to reduce air pollution and minimizing the associated health risks.

Additionally, the adoption of sustainable waste management practices, such as waste-to-energy systems, can utilize waste materials as fuel for combustion, reducing both waste volume and the need for fossil fuel consumption (Mukasyan, 2014).

6.6. COMBUSTION TECHNOLOGIES AND INNOVATIONS

In the quest for more efficient and environmentally friendly combustion processes, advanced combustion systems have emerged as promising solutions. These innovative technologies, including fluidized bed combustion, oxy-fuel combustion, and chemical looping combustion, are revolutionizing the way we approach combustion processes across various sectors.

Fluidized bed combustion is a highly efficient and flexible combustion technology that utilizes a bed of solid particles, typically sand or limestone, suspended in a stream of air or gas. This creates a fluidized state where the particles behave like a boiling liquid, allowing for efficient mixing and improved heat transfer. Fluidized bed combustion systems have found vast applications in power generation, industrial processes, and waste management. They offer advantages such as fuel flexibility, low emissions, and the ability to burn low-grade fuels, including biomass and waste materials (Pfefferle, 2012).

Oxy-fuel combustion is another advanced system that involves burning fuel in a mixture of oxygen and recycled flue gas instead of air. By eliminating the nitrogen present in air, oxy-fuel combustion enables the production of a fluid gas stream consisting primarily of CO_2 and water vapor, making carbon capture and storage (CCS) easier. The concentrated CO_2 stream can be captured and stored, mitigating greenhouse gas emissions. Oxy-fuel combustion is particularly suitable for applications where the combustion products need to be captured for carbon sequestration or utilization purposes, such as in power plants and industrial processes.

Chemical looping combustion (CLC) is a unique combustion process that utilizes metal oxide particles as oxygen carriers to facilitate combustion. In CLC, the fuel is reacted with a solid oxygen carrier, generating heat and transferring oxygen from the air to the fuel. This allows for the separation of CO_2 and other combustion products, simplifying their capture and enables more efficient carbon capture technologies. Chemical looping combustion has shown promise in power generation and is being explored for applications in industrial sectors such as steel production, where reducing emissions is a significant challenge.

These advanced combustion systems offer several benefits compared to traditional combustion technologies. They can significantly enhance combustion efficiency, resulting in higher energy conversion rates and lower fuel consumption. By minimizing air pollutants and greenhouse gas emissions, these technologies contribute to improve air quality and reduced environmental impact. Moreover, the use of alternative fuels, such as biomass and waste materials, in these systems helps reduce reliance on fossil fuels and promotes sustainable energy practices.

However, the implementation of advanced combustion systems also presents challenges. Development and scale-up of these technologies require significant research, engineering expertise, and investment. Integration with existing infrastructure and regulatory framework must be considered to ensure seamless adoption. Additionally, the availability and accessibility of alternative fuels play a crucial role in determining the viability and widespread adoption of these systems (Mukasyan, 2014).

6.7. SAFETY CONSIDERATIONS IN COMBUSTION REACTIONS

Combustion reactions are accompanied by inherent risks, one of the most significant being the potential for fires. It is crucial to prioritize fire safety measures and implement effective prevention strategies to minimize the likelihood of fires and their devastating consequences (Pfefferle, 2012).

6.7.1. Fire Safety

Fire safety begins with a comprehensive understanding of the properties of the materials involved in the combustion process. By assessing the flammability and ignition characteristics of substances, it becomes possible to identify potential fire hazards and to take appropriate precautions. This knowledge helps in selecting suitable fire-resistant materials and designing buildings with fire-resistant features. Fire-resistant materials play a vital role in reducing the spread and intensity of fires. These materials are specifically designed to withstand high temperatures and inhibit the progress of flames. Examples include fire-resistant coatings, fire-retardant fabrics, and fireproof insulation. By incorporating such materials into the construction of buildings, the risk of fire-related accidents can be significantly mitigated

Building design is another crucial aspect of fire safety. Structures should be designed with clear evacuation routes and adequate fire exits that facilitate swift and safe evacuation in case of a fire. Additionally, the installation of fire suppression systems, such as sprinklers and fire alarms, are imperative. These systems can detect fires at their early stages and either suppress them or alert occupants and emergency services promptly.

Fire prevention strategies play a pivotal role in minimizing the likelihood of fires. Regular inspection and maintenance of electrical systems, heating equipment, and other potential ignition sources are essential to identify and address any potential fire hazards. Furthermore, strict adherence to proper storage and handling of flammable materials, such as fuels and chemicals, is crucial to prevent accidental ignition.

Education and awareness are key elements in fire safety. Conducting training programs to educate individuals on fire prevention, emergency procedures, and the correct usage of fire extinguishers can greatly enhance preparedness and response capabilities. Encouraging a fire-conscious culture within workplaces and communities fosters a sense of responsibility and ensures that fire safety remains a priority.

6.7.2. Explosions and Hazards

Explosions are an extreme hazard associated with combustion processes, and understanding the conditions that lead to these events is of utmost importance in ensuring safety. Several factors can contribute to the occurrence of explosions, including the presence of flammable substances, ignition sources, and confinement (Mukasyan, 2014).

To prevent explosions, it is crucial to eliminate or control these contributing factors. This can be achieved by the implementation of stringent safety protocols, such as conducting thorough risk assessments, designing, and maintaining equipment to prevent leaks or malfunctions, and implementing effective ventilation systems to minimize the accumulation of flammable vapors (Pfefferle, 2012).

Risk assessment techniques are invaluable in identifying potential hazards and evaluating their severity. By analyzing the characteristics of the substances involved, the operating conditions, and the potential sources of ignition, it becomes possible to determine the likelihood and consequences of an explosion. This information enables the implementation of appropriate preventive measures and the development of emergency response plans.

Explosion prevention measures encompass various strategies, including the use of explosion-proof equipment and the implementation of effective control systems. Explosion-proof equipment is designed to contain potential explosions within enclosed spaces, minimizing their impact on the surroundings regarding hazards. Control systems, such as emergency shutdown procedures and interlocking mechanisms, can prevent catastrophic events by interrupting the combustion process or isolating hazardous areas.

6.7.3. Risk Assessment and Mitigation

Risk assessment and mitigation plays a crucial role in managing the potential risks associated with combustion reactions. These processes involve systematic evaluation, planning, and implementation of measures to minimize the likelihood and impact of accidents.

Risk assessment methodologies provide a structured approach to identify and assess potential hazards, evaluate the associated risks, and prioritize necessary actions. These methodologies typically involve in analyzing the properties and behavior of substances, considering process conditions, and assessing the likelihood of incidents occurring. Through this comprehensive evaluation, potential hazards can be identified and appropriate control measures can be implemented (Mukasyan, 2014).

Safety regulations and guidelines serve as important frameworks for ensuring adequate risk management in combustion processes. Regulatory bodies establish standards and requirements that govern the design, operation, and maintenance of facilities to mitigate potential risks. Compliance with these regulations is essential to maintain a safe working environment and prevent accidents.

Developing and implementing emergency response plan is crucial for effectively handling incidents and minimizing their impact. These plans outline the necessary actions to be taken in the event of a fire, explosion, or other emergencies. They include protocols for evacuations, communication channels, emergency contact information, and procedures for containing and extinguishing fires. Regular drills and training sessions should be conducted to ensure that personnel are well-prepared to execute the emergency response plans effectively.

6.8. POLICY AND REGULATION FOR COMBUSTION CONTROL

Policy and regulation plays a vital role in managing the environmental and safety risks associated with combustion. Governments, international organizations, and industry stakeholders all have a significant responsibility in developing and implementing effective regulations that promote cleaner and safer combustion practices (Pfefferle, 2012).

Governments have a crucial role in setting standards and regulations to control combustion-related emissions and ensure the safety of their citizens. They can establish emission limits for different industries and sectors, such as transportation, power generation, and manufacturing. By enforcing these limits, governments can encourage the adoption of cleaner technologies and fuels, leading to reduced environmental impact. Additionally, they can implement safety regulations to prevent accidents and mitigate the risks associated with combustion processes, ensuring the well-being of workers and the general public (Mukasyan, 2014).

International organizations also play a crucial role in developing policies and regulations that transcend national boundaries. Organizations such as the United Nations Environment Program (UNEP) and the World Health Organization (WHO) can provide guidance and facilitate international cooperation in addressing combustion-related issues. Through global agreements and initiatives, these organizations can encourage countries to harmonize their regulations, share best practices, and collaborate on research and development efforts to advance cleaner and safer combustion technologies.

Industry stakeholders, including manufacturers, energy providers, and technology developers, have a major role to play in shaping combustion regulations. These stakeholders can contribute by proactively adopting cleaner and more efficient technologies, investing in research and development, and participating in policy discussions. By engaging with policymakers, they can provide valuable insights based on their expertise and contribute to the development of practical and effective regulations. Collaboration between industry and regulatory bodies is essential to ensure that regulations are both feasible and impactful, taking into account technological advancements and economic considerations.

In developing policies and regulations for combustion control, it is crucial to consider a holistic approach that balances environmental, safety, and economic factors. Regulations should be based on scientific evidence and consider the specific characteristics and challenges of different industries and sectors. Flexibility is key, as it allows for continuous improvement and adaptation to new technologies and knowledge. Moreover, policies should incentivize innovation and research in combustion technologies, as ongoing advancements can lead to significant improvements in efficiency, emissions reduction, and safety (Pfefferle, 2012).

CHAPTER

7 Acid–Base Neutralization Reactions

Contents

An acid and base react to form a hydroxide ion (OH). Water-based acids produce H+. Arrhenius established acid-base definitions in 1884. H^+ ions can represent unpaired hydrogen atoms or protons. Aqueous proton mobility? No. H+ and water molecules form covalent connections regularly. H_3O^+ is a protonated water molecule. Hydronium ions generate hydrogen ions in aqueous solutions because of their rationality (Sun, 2012) (Figure 7.1).

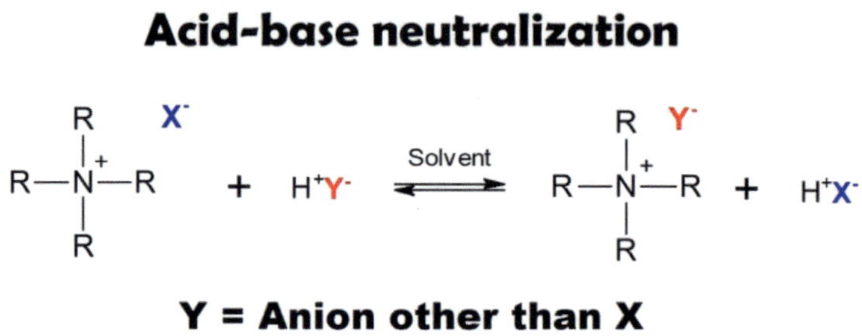

Figure 7.1. *Acid-base neutralization reaction.*

Source: Ebrahimi, M., Kujawski, W., Fatyeyeva, K., & Kujawa, J., (2021). A review on ionic liquids-based membranes for middle and high temperature polymer electrolyte membrane fuel cells (PEM FCs). International Journal of Molecular Sciences, 22, 5430. – Scientific Figure on ResearchGate. Available from: https://www.researchgate.net/figure/Synthesis-of-secondary-ILs-via-acid-base-neutralization-reaction-Reprinted-with_fig3_351773833.

7.1. ACID–BASE DEFINITIONS

In 1754, Guillaume-François Rouelle used the term "base" to designate a substance that reacts with an acid to form a salt. Most bases are bitter.

1. **Lavoisier's Oxygen Theory of Acids:** In 1776, Lavoisier made acids and bases known to science. Lavoisier defined acids in terms of their oxygen content, giving them names derived from Greek words that mean "acid-former" (from the Greek (oxys) meaning "acid"). His knowledge of strong acids was limited to oxoacids, such as HNO_3 (nitric acid) and H_2SO_4 (sulfuric acid), which tend to contain central atoms in high oxidation states surrounded by oxygen (McLuckey, 2013).

   Before Sir Humphry Davy's 1810 article and lectures, which demonstrated the absence of oxygen in H_2S, H_2Te, and hydrohalic acids, the Lavoisier definition was accepted for over 30 years. By concluding that "acidity does not depend upon any particular elementary substance, but upon the particular arrangement of various substances," Davy was unable to develop a novel hypothesis. By proposing that acids are nonmetal oxides and bases are metal oxides, Jöns Jacob Berzelius amended the oxygen hypothesis. Antoine-Laurent Lavoisier initially sought to logically describe acid behavior towards the end of the 18th century. Lavoisier proposed that all acids must contain oxygen, which was reflected in the names given to this element in various languages, such as the English oxygen, which comes from the Greek words *oxys* (sour) and *Genna* (production), the German Sauerstoff, which means "acid material," and the Russian kislorod (Sun, 2012).

2. **Liebig's Hydrogen Theory of Acids:** In 1838, Justus von Liebig proposed that an acid is a molecule containing hydrogen that may be replaced by a metal. Davy's theological transition from oxygen-based to hydrogen-based acids was completed by his important study on the chemical makeup of organic acids. Before Arrhenius' concept was established, Liebig's empirical definition was in use for more than 50 years. Hydrochloric acid was oxygen-free, enabling this. German scientist Justus von Liebig defined an acid in 1838. Liebig defined an acid as a chemical that permits the metal to replace hydrogen. For nearly 50 years, this definition ruled the field. Liebig's theory assumed that bases formed ions when they neutralized acids, but nothing was known about their constitutional features.

3. **Arrhenius Definition:** Acids and bases were first defined molecularly by Svante Arrhenius. For his hydrogen hypothesis of acids, which came after his 1884 work on ions in aqueous solution with Friedrich Wilhelm Ostwald, Arrhenius won the Nobel Prize in Chemistry in 1903. A substance that dissociates in water to produce hydrogen ions (H^+) is what Arrhenius described as an acid.

   To form the hydronium (H_3O^+) ion, water protonates. H^+ is now used as a shorthand for H_3O^+ since aqueous solutions do not contain free protons. This species is measured using pH indicators.

   The concentration of hydroxide (OH^-) ions in an aqueous solution rises as a result of an Arrhenius base dissociating in water to produce these ions. The Arrhenius definitions of acidity and alkalinity, which relate to solvent ion concentration, cannot be applied to the majority of non-aqueous solutions. While molten NaOH and calcium amide in liquid ammonia are not alkaline, pure H_2SO_4 and HCl in toluene are not acidic. Brnsted-Lowry and Lewis theories were established for non-aqueous exceptions. An acid-base reaction is neutralization. Both salt and water are produced (McLuckey, 2013).

4. **Hydrogen-Hydroxide Ions:** The electrolytic dissociation hypothesis, developed by Nobel laureates Wilhelm Ostwald and Svante August Arrhenius in the 1880s, quantified acid-base chemistry. This idea states that electrolytes dissolve into ions. Only hydrogen compounds that produce hydrogen ions (H^+) in aqueous solution are acids. As shown by vegetable colors and other features, hydrogen ions in a solution correspond with acidity. Thus, acid neutralization by bases may be described as how these two ions react with each other to generate water (($H^+ + OH^- \rightarrow H_2O$), and basic (or alkaline) qualities can be connected with the concentration of hydroxide ions (OH^-) in an aqueous solution. This naturally defines acids and bases as compounds that create hydrogen and hydroxide ions in aqueous solution. This definition was popular for 30–40 years. Its notion of bases is reasonable, although Liebig's definition of acids is purer (Sun, 2012).

However, quantifying acids and bases in terms of hydrogen and hydroxide ions have advantages. Due to the solution's electrical conductivity, hydrogen, and hydroxide ions may be quantified, indicating its acidity or alkalinity. Reversible process equations may also determine acid and base dissociation constants. These constants quantify acid and base strengths (degrees of dissociation), replacing semi quantitative approximations. This technique quantitatively explained many experimental data, a description that has survived later acid-base reaction definition developments.

These quantitative gains disguised substantial uncertainties and logical contradictions in the qualitative definitions of acids and bases when forming hydrogen and hydroxide ions. Anhydrous hydrogen chloride, which does not conduct electricity, was unclear whether it was an acid before or after water contact. To apply the definition of bases to nonaqueous solutions, where the anion (negatively charged ion) produced depends on the solvent, such as the methoxide ion (CH_3O^-) in methanol (CH_3OH) and the amide ion (NH_2^-) in liquid ammonia (NH_3), the definition had to be revised. The solvent is also implicated in acids since the so-called hydrogen ion in solution is always H_3O^+ in water, CH_3OH^{2+} in methanol, and NH^{4+} in liquid ammonia. These qualities inspired solvent-dependent acid and base definitions. Despite this discovery, common acid-base activities including neutralization, indicator (vegetable pigment) effects, and catalysis still occurred in solvents like benzene or chloroform where conductivity tests could not identify free ions. In aqueous solutions, bases are complicated because some, like metallic hydroxides, have hydroxyl groups while others, like amines, do not. These produce hydroxide ions with water molecules in the solution (McLuckey, 2013).

J. M. Brnsted and T. M. Lowry developed a new and more extensive definition of acids and bases nearly simultaneously in 1923 to fix the many flaws in the hydrogen-hydroxide ion definitions. The Bronsted-Lowry definition of acids and bases has had a major impact on the understanding of many phenomena and has inspired a lot of experimental work, even though precise verbal definitions of qualitative concepts are usually unprofitable in physical science. Defined below: Acids lose protons, whereas bases acquire them. The definition of a proton is solvent-independent since it refers to H+, the nucleus of a hydrogen atom, not hydrogen ions. Instead of substance or molecule, species implies that acid and base apply to positively and negatively charged ions and uncharged molecules. The Bronsted-Lowry definition relies on this extension. Conjugated acid-base pairs A and B symbolize it. This combination has no sign or magnitude constraints on the charges, save that A must have one more positive charge (or one fewer negative charge) than B (Sun, 2012).

Other definitions of acids and bases as proton donors and acceptors are used, but Bronsted-Lowry is the most common. The most important concept is more generic than the Bronsted-Lowry definition, while others are unique to certain situations. In 1923, scientist Gilbert N. Lewis invented this phrase.

Lewis described an acid as a substance that absorbs an electron pair from a base to form a covalent bond. This categorization contains the same species as the Bronsted-Lowry definition of bases since a molecule or ion with one or more unshared pairs of electrons can take a proton. Thus, it can combine with electron and proton acceptors. Due to their extended outer electron shells, boron trifluoride (BF_3), sulfur trioxide (SO_3), and silver ion (Ag^+). Do not contain protons. Experimental data corroborated Lewis' early beliefs that non protonic acids had acidic qualities like neutralizing bases, acting on indicators, and catalyzing chemical processes. This is not always true; CO_2 and nitrogen pentoxide (N_2O_5) have entire octets of electrons and cannot accept more electrons according to traditional valence theory. Because they are adducts of the proton (a real Lewis acid) and a base, such as an acetate ion, hydrogen-containing acids like acetic acid are not electron acceptors. The Lewis scheme can only include proton acids, represented by XH, and bases, represented by B if they transit through an intermediate hydrogen-bonded state and a weak secondary attractive force (McLuckey, 2013).

Lewis and Bronsted-Lowry's definitions have been debated extensively. Terminological distinctions have a minimal scientific value. In Bronsted-Lowry nomenclature, acids are proton donors and Lewis's acids are electron-pair acceptors. Proton acids simplify the quantitative description of acid-base processes (see below, Acid-base equilibria). The most common (Sun, 2012).

7.1.1. Proton-Transfers

Bronsted-Lowry does not include $A \rightleftarrows B + H^+$ in solution procedures. The H^+ ion, commonly known as the naked proton, is only visible at very low pressure yet is essential to the composition of most chemical entities (Figure 7.2).

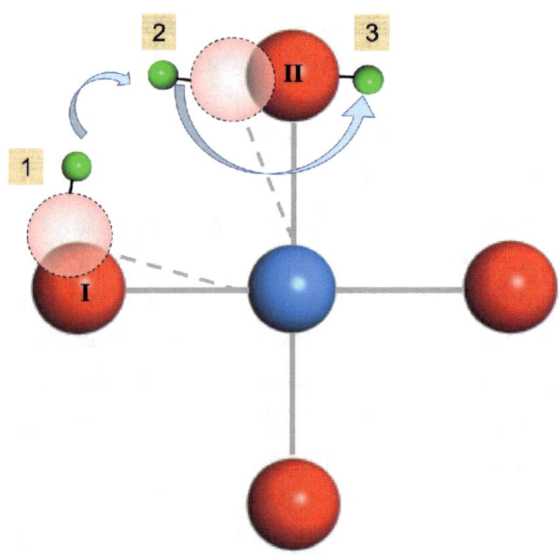

Figure 7.2. *Proton transfers.*

Source: Meng, Y., Gao, J., Zhao, Z., Amoroso, J., Tong, J., & Brinkman, K.,

(2019). Review: Recent progress in low-temperature proton-conducting ceramics. Journal of Materials Science. – Scientific Figure on ResearchGate. Available from: https://www.researchgate.net/figure/Schematic-illustration-of-possible-proton-transfer-mechanism-in-perovskite-type_fig2_332340555.

Due to its small size of (about 10^{-15} meter), the proton's tremendous electric field may polarize and attracts any molecule or ion. Dissociating 19 grams of H_3O^+ hydronium ion into 1 gram of protons and 18 grams of water using 1,200,000 joules (2,90,000 calories) seems implausible. Acid-base reactions have two reaction pathways:

$A_1 \rightleftarrows B_1 + H^+$ and $H^+ + B_2 \rightleftarrows A_2$, leads to $A_1 + B_2 \rightleftarrows B_1 + A_2$.

This shows many solution reactions and their nomenclatures. A_1, B_2, B_1, and A_2 are taxonomic nomenclatures.

1. **Dissociation of Molecular Acids in Water:** Water underpins. Acetic acid dissociates as:

   $CH_3CO_2H + H_2O \rightleftarrows CH_3CO_2^- + H_3O^+$.

2. **Dissociation of Bases in Water:** Water donates a proton to the base as an acid:

   $H_2O + NH_3 \rightleftarrows OH^- + NH_4^+$

3. **Dissociation of Acids and Bases in Non-Aqueous Solvents:** Aqueous solution reactions are comparable. Ammonia decomposes into $CH_3CO_2H + CH_3OH \rightleftarrows CH_3CO_2^- + CH_3OH$ while acetic acid decomposes as:

   $CH_3OH + NH_3 \rightleftarrows CH_3O^- + NH_4^+$.

4. **Self-Dissociation of Amphoteric Solvents:** The solvent has basic and acidic molecules. Water-ammonia phase separation.

7.2. ACID BASE NEUTRALIZATION REACTIONS

Neutralization is a chemical reaction in which equal concentrations of acid and base react. In water processes, neutralization removes hydrogen and hydroxide ions. The pH of the neutralized solution is determined by the acidity of the reactant. In chemistry, an acid-base reaction is referred to as neutralization. Historically, this reaction was represented as

Acid + base (alkali) $\rightarrow$ salt + water

Examples include:

$NaCl + H_2O = HCl + NaOH$

The assumption is admissible as long as it is understood that dissociation alters the ionization state of chemicals in aqueous solutions. Because neutralization is a quantitative process, the arrow sign is used. A larger concept is provided by Brnsted-Lowry's acid-base theory (Sun, 2012).

$A+BH = AH+B$

Electrical charges are not included in generic terms like these since each species A, AH, B, or BH may have one. Sulfuric acid neutralization is one example. There are two possible partial neutralization reactions.

$H_2SO_4 + OH^- \rightarrow HSO^-_4 + H_2OHSO^-_4 + OH^- \rightarrow SO^{2-}_4 + H_2O$

Overall: $H_2SO_4 + 2\ OH^- \rightarrow SO^{2-}_4 + 2\ H_2O$

When an acid AH is neutralized, it eliminates all molecules and hydrogen ions from the solution. To neutralize an acid, the base must equal the acid. This foundation is comparable. In an acid-base titration, the comparable point is neutralization. Acid and alkali concentrations are used to quantify the neutralization process. At the point of equivalence:

Volume (acid) x concentration (dissociated H+ ions) equals volume (base) x concentration (OH) (McLuckey, 2013).

In general, the volumes of an acid AH_n interacting with a base $B(OH)_m$ at concentration c_2 are as follows:

$n\, v_1\, c_1 = m\, v_2\, c_2$

AN example

Bases are neutralized by acids.

$Ba(OH)_2 + 2\ H^+ \rightarrow Ba^{2+} + 2\ H_2O$

The acid-base equation is applicable. Neutralization is not limited to solution reactions. To neutralize, limestone interacts with sulfuric acid.

$[Ca, Mg]CO_3(s) + H_2SO_4(aq) \rightarrow (Ca^{2+}, Mg^{2+})(aq) + SO^{2-}_4(aq) + CO_2(g) + H_2O$

Such reactions are required in soil chemistry.

7.2.1. Strong Acids and Bases

In water, strong acids completely dissociate. HCl is a strong acid.

$HCl(aq) \rightarrow H^+(aq) + Cl^-(aq)$

In water, strong bases completely dissociate. NaOH is a strong base (Sun, 2012).

$NaOH(aq) \rightarrow Na^+(aq) + OH^-(aq)$

As a result, a strong acid interacting with a strong base neutralizes as follows:

$H^+ + OH^- \rightarrow H_2O$

Without the presence of Na^+ and Cl^- ions, hydrochloric acid, and sodium hydroxide react. Because the hydrogen ion exists as the hydronium ion, the neutralization process may be described as follows:

$H_3O^+ + OH^- \rightarrow H_2O + H_2O$

Strong bases neutralize strong acids without producing more hydrogen ions. Neutral solutions have neither acidic nor alkaline properties. The pH of such a solution is close to 7, however, it varies depends on the temperature. Neutralization using exothermic means (McLuckey, 2013).

$H^+ + OH^- \rightarrow H_2O$ is −57.30 kJ/mol.

7.2.2. Quantitative Treatment

When the concentration of an undissociated solute falls below the detection limits, the solute is considered entirely dissociated and cannot be measured. In terms of numbers, log K 2 or 1.76. Experiments cannot be used to calculate the dissociation constant. It is theoretically possible to estimate. The log K 6 of hydrogen chloride in an aqueous solution at room temperature has been predicted. At low concentrations, a chemical compound can operate as a strong acid in solution and as a weak acid at high concentrations. As in sulfuric acid (Sun, 2012).

In water, a weak acid, HA, partly dissociates. As a result, an equilibrium mixture is formed:

$H_3O^+ + A^-$ gives H_2O + HA

Acetic acid is a weak acid. Its pH. The acid dissociation constant, Ka, is used in HA + OH gives H_2O + A. In titration, calculate the end-point or equivalence point pH. Finally, the acid is neutralized, so the analytical hydrogen ion concentration, TH, is zero and the conjugate base concentration, A, equals the analytical or formal concentration of the acid, TA. The concentrations of HA solutions in equilibrium are connected by the equation

log Ka [A] = pKa[H+]

Acid + base = salt + water

Acid-base neutralization is illustrated with a double-replacement process.

A solution of sodium chloride and water is created when hydrochloric acid (HCl) and sodium hydroxide (NaOH) combine.

$$NaCl(aq) + H_2O = HCl(aq) + NaOH(aq)$$

In this equation, Arrhenius included (aq). Chemicals are dissolved in water. In aqueous solutions, the pure chemicals HCl, NaOH, and NaCl dissociate into H^+, Cl, Na, and OH.

Water reacts with sodium bicarbonate and acid salts to produce CO_2. No matter if the baking powder is produced or bought, the compositions are the same. This is an example of an acid-base reaction:

$$NaHCO_3 + H^+ \rightarrow Na^+ + CO_2 + H_2O$$

Acids make reactions harder. CO_2 is produced using the following stoichiometry using monocalcium phosphate and baking soda:

$$14\ NaHCO_3 + 5\ Ca(H_2PO_4)_2 \rightarrow 14\ CO_2 + Ca_5(PO_4)_3OH + 7\ Na_2HPO_4 + 13\ H_2O$$

Monocalcium phosphate ("MCP") is a component in baking powders. Common weight-based formulas include 30% sodium bicarbonate, 5 to 12% monocalcium phosphate, and 21 to 26% sodium aluminum sulfate. Commercial baking powder may use sodium acid pyrophosphate rather than sodium aluminum sulfate. Another tartaric acid derivative utilized in these formulations is a cream of tartar ($KC_4H_5O_6$).

Acids neutralize bases. Water is safe when acids and bases neutralize. Salts can be soluble or insoluble when acids and bases react. In chemistry, "salt" includes more than table salt. HCl(aq)+ KOH (aq)The reaction between liquid H_2O and aqueous KCl is one example. KCl is salt. An element's atoms contain one water molecule. Adding HCl and H_2O molecules to the chemical reaction between HCl(aq) and $Mg(OH)_2$ (aq) maintains equilibrium (Sun, 2012).

$$2HCl(aq)+Mg(OH)_2(aq)\rightarrow 2H_2O(\ell)+MgCl_2(aq)2HCl(aq)+Mg(OH)_2(aq)\rightarrow 2\ H_2O(\ell)+MgCl_2(aq)$$

This dish requires $MgCl_2$ salt. Base-containing antacids reduce stomach acid. Hydronium and hydroxide ions are equivalent in neutral solutions. Acid-base neutralization occurs. Equimolar amounts cannot neutralize in stoichiometry. Salt type affects a solution's acidity, neutrality, or basicity. Four examples show how neutralization with stoichiometrically equivalent numbers can produce solutions with different pH levels:

HCl(aq) and NaOH(aq) neutralize because their conjugate partners are weak, form a neutral solution.

$$HCl(aq)+NaOH(aq)\rightleftharpoons NaCl(aq)+H_2O(l)(7.4.1)\ (7.4.1)HCl(aq)+NaOH(aq)\rightleftharpoons NaCl(aq)+H_2O(l)$$

A weak base and strong acid react to generate a slightly acidic solution and a conjugate acid. A strong base and weak acid solution are moderately basic. Strong bases and weak acids react to form conjugate bases and acids. The solution pH remains unchanged because the conjugate acid of the strong base is less acidic than water. Deprotonated weak acids seldom dissociate in an aqueous solution. Hydrogen ion generation increases solution basicity. Weak acid-base solutions can be acidic, basic, or neutral. This solution is the most difficult of the four. The conjugation's acid-base strengths determine a solution's acidity or basicity. A neutral pH might occur from equal conjugate base and acid concentrations. Conjugates have acid-base equimolarity. Conjugate K values dictate acidity, basicity, or neutrality (Sun, 2012).

Neutralizing reactions have these rules:

- Double displacement redox reactions include neutralization reactions.
- Remember displacement response articulation guidelines.
- Choose reactants and products.
- Double replacement reactions swap cations.
- Single replacement reactions produce a new product when the element replaces the cation in the reacting molecule.
- Check each chemical formula for oxidation state.

Write the balanced chemical equation for the neutralization of HCl with $Mg(OH)_2$. Hint: neutralization reactions are a specialized type of double replacement reaction.

$$2HCl + Mg(OH)_2 \rightarrow Mg(Cl)_2 + 2H_2O$$

Since HCl is a strong acid and $Mg(OH)_2$ is a strong base, the resulting solution would be neutral which is a mixture of salt and water.

7.2.3. Stomach Antacids

0.03 M HCl improves food processing in the intestines. Reflux causes burning because a muscular valve at the top of the stomach lets gastric acid into the lower esophagus. The esophagus is not protected from gastric acid like the stomach; therefore, the effects can be terrible. Antacids lower esophageal acid and relieve reflux. Antacids are bases. CaCO3 is a popular antacid. $CaCO_3(s)+2HCl(aq)$ gives $CaCl_2(aq)+H_2O(l)+CO_2(g)$……. Reaction (7.4.2). $CO_2(g)$ neutralizes stomach acid and may generate a satisfying belch. Magnesium-containing antacids may be cathartic even without CO_2. $Al(OH)_3$ is an active ingredient in several antacids. Anti-constipation antacids contain magnesium and aluminum hydroxides (McLuckey, 2013).

7.2.4. Culinary Aspects of Chemistry

Cooking is safe-to-eat synthetic chemistry. Sustenance uses acid-base chemistry often. Baking soda (sodium bicarbonate) is an example. Primary $NaHCO_3$. CO_2 gas pockets form when carbonic acid combines with an acid in a batter, such as lemon juice, buttermilk, or sour cream, and the batter "rises." Baking powder is created when sodium bicarbonate and acid chemicals combine with batter water.

Many people put acidic vinegar or lemon juice on their cooked fish. Studies show that fish have volatile amines (bases) that acids neutralize to generate involatile ammonium ions. This diminishes the fish's fragrance and adds a "sour" taste we like. An acid-base reaction exchanges hydrogen ions, H^+, between neutral (water, H_2O, or acetic acid, CH_3CO_2H) or electrically charged (ions, such as ammonium, NH_4^+; hydroxide, OH^-; or carbonate, CO_3^{2-}) species. Aluminum chloride, $AlCl_3$, and the silver ion Ag^+ behave similarly.

Acids damage metals that taste sour and turn blue vegetables red. Bases turn red vegetable pigments blue and feel soapy when mixed with water. Acids and bases form salts, which have a salty taste but no acidic or basic characteristics. Acids and bases form salts (Sun, 2012).

Medieval alchemical works introduced the terminology "acid," "base," and "salt" to describe compounds as acids or bases. Acids are likely identified originally because of their taste. Acid, French acide, German Saure, and Russian kislota come from Latin acidus, German sauer, Old Norse sur, and Russian kisly. Acids were first noted for their solvent or corrosive qualities, their influence on vegetable colors, and their CO_2 clouds when applied to chalk. Bases (or alkalies) neutralized acids and formed salts and water, which were usually crystalline, water-soluble, and saline-smelling.

Despite their ambiguity, these theories explained a wide range of qualitative observations, and many of the most common chemical substances encountered by early chemists were acids (hydrochloric, sulfuric, nitric, and carbonic acids), bases (soda, potash, lime, ammonia), or salts (common salt, ammoniac, saltpeter, alum, and borax). Because acid-base behavior had no physical foundation, quantifying it was challenging. However, the capacity of a constant quantity of acid to neutralize a fixed amount of base was one of the first demonstrations of chemical equivalence, the idea that one substance is chemically identical to another. Early on, it was discovered that one acid in a pinch of salt may be substituted by another, allowing acids to be approximately sorted by potency. It was obvious that many of these displacements might go either way depending on the testing settings. This showed that acid-base processes can replenish the initial material. It also promoted acid-base balance. This theory states that reversible chemical processes approach equilibrium when one reaction can regenerate both the starting materials and the products as rapidly as the other (McLuckey, 2013).

Acids and bases are essential in industrial chemistry, daily life, and theory. Sulfuric acid and sodium hydroxide, two of the most produced compounds, are reactants or catalysts in many chemical reactions. Plants and animals depend on soil and water acidity or alkalinity, which is connected to almost every biological chemical activity. Acid-base chemistry and the word "salt" are everywhere in modern life. $H_3O^+ + OH^- \rightleftarrows 2H_2O$ neutralizes strong acids and bases in water. Weak acids and bases neutralize by directly transferring protons. Acetic acid and ammonia combine to form as $CH_3CO_2H + NH_3 \rightarrow CH_3CO_2^- + NH_4^+$. The neutralization equation shows neutralization with or without benzene. If one reactant is in excess, the process can be described as the dissociation of ammonia in glacial acetic acid or liquid ammonia (Sun, 2012).

7.3. HYDROLYSIS OF SALTS

Many chemical substances create acidic or basic solutions. Previous acid-base model hydrolysis processes were difficult. According to Bronsted-Lowry's theory, mild acids and bases form cations and anions with acidic and basic characteristics, which enable hydrolysis. Due to their low interaction, sodium-chloride equations can be ignored. The hydronium ion (H_3O^+) gives the first equation acidity, whereas the hydroxide ion (OH^-) gives the second equation basicity. A powerful acid neutralizes aqueous ammonia, yet a forceful base neutralizes acetic acid. Aqueous solutions hydrolyze metal salts, especially those with multivalent cations. Chromic, ferric, and aluminum salts produce acidic solutions. Proton-transfer processes cause highly hydrated ions. Protons move from hydration shell water molecules to solvent water molecules in aluminum salt solutions. Hydronium ions (H_3O^+) cause acidity (Figure 7.3).

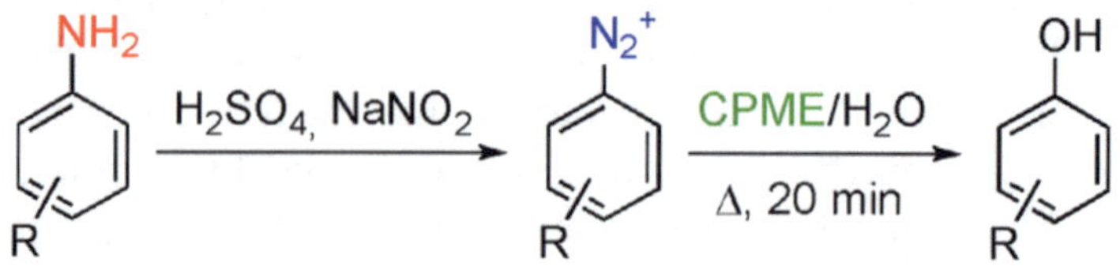

R = H, 3-(4-NO_2-C_6H_4O), 3-C_6H_5O, 3-$C_6H_5CH_2$O, 3-*i*Pr; 3-Me; 3-MeO; 3-Ac; 3-NO_2, 3-CF_3, 3-COOH, 2-Me-4-COOH, 4-Me-2-COOH; yield: 56-98%.

Figure 7.3. *Hydrolysis of diazonium salt in CPME/H_2O.*

Source: Azzena, U., Carraro, M., Pisano, L., Monticelli, S., Bartolotta, R., & Pace, V., (2018). Cyclopentyl Methyl Ether: An Elective Eco-Friendly Ethereal Solvent in Classical and Modern Organic Chemistry. – Scientific Figure on ResearchGate. Available from: https://www.researchgate.net/figure/Scheme-63-Hydrolysis-of-diazonium-salt-in-CPME-H-2-O_fig1_327843196.

7.4. ACID–BASE CATALYSIS

Bases and acids, especially, Lewis's acids, accelerate many chemical processes in laboratories, businesses, and nature. Catalytic activity and electrical conductivity in the late 19th century led to the discovery of electrolytic dissociation as the main mechanism affecting acid-base behavior.

The current belief is that acid-base catalysis involves an actual acid-base interaction between the catalyst and substrate, which regenerates the catalyst during the reaction. As reaction processes are more understood, many scientists proposed event sequences for acid- or base-catalyzed reactions are realistic and often established (McLuckey, 2013).

Protons do not affect molecule structure, stability, or reactivity in most acid-base processes. A substrate's reactivity or instability increases with proton addition or removal. Acids or bases stimulate their disintegration or rearrangement. Protons are often added or removed during catalyst renewal during rearrangement. Small acid-base interactions can alter acid- or base-catalyzed processes (Sun, 2012).

7.5. ACID–BASE EQUILIBRIA

Due to their acidic and basic qualities, water, alcohols, ammonia, amines, and acetic acid behave differently. Lyonium and lyate ions can be formed by transferring a proton from the solvent molecule SH. These ions alone keep purified solutions electrically neutral. The chemical equation $2SH = SH^{2+}$, $+ S^-$ represents equilibrium. The equilibrium constant (Ks′) for this reaction is calculated using the formula $Ks' = [(SH_{2+}),] [S^-].]/[SH]^2$, where square brackets indicate species concentrations. Since the solvent concentration, [SH], is constant and important, it is usual to remove this term and represent the solvent's self-dissociation as $Ks = [(SH_{2+}),][S]^2$. Ks is the solvent ion product or auto-protolysis constant. Concentrations in moles per liter are based on the molecular weight of a substance in grams. The solvent's dual function as a proton donor and acceptor lowers the ionization degree and Ks. Sulfuric acid's greatest Ks value at full concentration at 25°C is 1.7×10^{-4}. The equation and the solvent apply to diluted solutions of acids, bases, or salts with the same Ks. Excess ions in a solution imply electrical neutrality, therefore $[(SH_{2+})]$ and [S] need not be comparable. The equation shows that low [S] values mean high $[(SH_{2+})]$ values and vice versa.

Acid A and solvent SH combine to create product $A + SH \rightleftarrows B + SH_2^+$, The equilibrium constant $[B][SH_2^+]/[A]$ [SH] describes the phenomenon. The equilibrium constant Ka is calculated using the constant dissociation of acid A in SH and the constant solvent concentration [SH]. If A1 is much stronger than A2, the acid-base interaction of A1 + B2 will shift left. Using the equilibrium constant to compare A1 and A2 potencies are reasonable. The dissociation constant measures chemical A's acid strength relative to the lyonium ion (SH_2^+), ignoring the constant component [SH]. Estimating the equilibrium constant when the reaction switches from left to right is difficult. SH's strong acid is A. Weak acids have measurable dissociation constants of less than 0.1 (Sun, 2012).

7.6. AQUEOUS SOLUTIONS

Aqueous solutions are essential to laboratory studies and flora and fauna's biological activities; hence they must be handled carefully and separately. Kw, the product of hydronium and hydroxide ion concentrations, is temperature-dependent. Kw is 1.0×10^{-14} mole^2 liter^{-2} at 25°C, 1.0×10^{-15} at 0°C and 7×10^{-13} at 100°C. Kw may be determined from the electrical conductance of purified water with $[H_3O^+] = [OH^-] = 10^{-7}$ at 25°C. Its manufacturing requires salt hydrolysis and other methods. Modifying the acid-base ratio creates a buffer. Mixing acetic acid with sodium acetate, partial neutralization with sodium hydroxide, or adding less than one equivalent of a strong acid to sodium acetate produces buffered solutions with different acetic acid and acetate ion concentrations. Ammonia, ammonium salt, or less than one equivalent of sodium hydroxide can be used to create a buffer system based on the NH_4^+-NH3 ion pair. The typical buffer solution hydrogen ion concentration formula is utilized (McLuckey, 2013).

Due to their frequent sub-unity values and large range, hydrogen ion concentration negative logarithms are useful. The negative logarithm represents real concentrations, which are usually between –1 and 13. With a base of 10, pH is proportional to the logarithm of hydronium ion (H_3O^+) concentration. Since hydronium ions ($[H_3O^+]$) are 1×10^{-7}, distilled water has a pH of 7. Alkaline solutions have a pH of 13.0, $[OH^-] = 0.1$, $[H_3O^+] = K_w/[OH^-] = 1 \times 10^{-14}/0.1 = 10^{-13}$,

The buffer ratio from solution preparation implements pH in buffered solutions. The buffer ratio and acid pKa define the buffered solution's pH. The phenomena are unaffected by A and B concentrations. Dilution is little affects the buffered solution pH. The system is insensitive to acid or alkali concentrations below [A] and [B]. An acid-base system's buffering capability is best between $pH = pK_a + 1$ to $pH = pK_a - 1$, or buffer ratios of 0.1 to 10. A or B declines too much to reduce the buffering effect (Sun, 2012).

7.6.1. Nonaqueous Solvents

Despite water's unusual solvent characteristics, aqueous solutions' acid-base properties have been studied due to their practical importance. This liquid may act as an acid and a base and has a high dielectric constant. This discussion will summarize acids and bases in various solvents. Acidic or basic solvents affect acid-base dissociation. Solvent dielectric constants that increase ion production affect several acid-base processes. The proton that forms SH_2^+ is fundamentally connected to the solvent's basicity, however, other ions may have a high solvation impact, interacting closely with the solvent. The dielectric constant and solvation effects overlap. Several popular solvents demonstrate these principles. Solvents can be amphoteric, acidic, basic, or aprotic. Mixed solvents include concentrated aqueous acids.

7.6.2. Amphoteric Solvents

Methanol and ethanol dominate this group of nonaqueous solvents. These substances have acid-base properties like water, but their dielectric constants are lower, reducing ion production efficiency. The solvents have four- to five-order-of-ten lower acid and base dissociation constants and ion products. $K_s = 10^{-17}$ for CH_3OH and 10^{-19} for C_2H_5OH, which are lower than 10^{-14} for water). As seen by its Ka value of 20, most nitric acid dissociates in water. Methanol has 2.5×10^{-4} Ka. Since the three solvents cannot change the ion count, their equilibrium constants are uniform (McLuckey, 2013).

7.6.3. Acidic Solvents

The most important extremely acidic solvent is sulfuric acid, which can protonate a broad range of oxygen- or nitrogen-containing compounds. Consequently, water, alcohols, ethers, ketones, nitro compounds, and sulfones are bases of sulfuric acid. Due to its potent ionic product high ($[H_3SO_4^+]$ $[HSO_4^-] = 1.7 \times 10^{-4}$), this solvent must have some fundamental characteristics. However, the exceptionally high acidity of the solvent conceals its basicity. Carboxylic acids, for instance, almost completely react with sulfuric acid according to the equation $RCO_2H + H_2SO_4 \rightarrow RCO_2H_2^+ + HSO_4^-$, acting as strong bases. Several compounds in sulfuric acid endure more complex reactions than simple proton transfers, typically resulting in notable chemically reactive species. Some alcohols, such as triphenylcarbinol $(C_6H_5)_3COH + 2H_2SO_4 \rightarrow (C_6H_5)_3\ C^+ + H_3O^+ + 2HSO_4^-$, react with sulfuric acid to produce carbonium ions. Nitric acid generates the nitronium ion, NO_2^+, through the reaction of $HNO_3^+\ 2H_2SO_4 \rightarrow NO_2^+ + H_3O^+ + 2HSO_4^-$ this ion is frequently employed as the catalyst in the nitration of organic molecules. Similar to sulfuric acid in terms of solvent properties, hydrogen fluoride is less acidic and less basic than sulfuric acid. Another acidic solvent that has been extensively studied is acetic acid. Due to its low dielectric constant, ions in it typically exist as ion pairs; however, more complex relationships are frequently formed. Consequently, quantifying acid-base equilibria in acetic acid can be challenging, although generalizations are possible. According to $B + CH_3CO_2H \rightarrow BH^+ + CH_3CO_2^-$. All aqueous chemicals more basic than aniline react completely with acetic acid. Consequently, all of these bases form solutions with identical acid-base properties; this is known as the leveling effect of the solvent. For acids, the opposite holds. For instance, the strong mineral acids nitric, hydrochloric, sulfuric, hydrobromic, and perchloric (HNO_3, HCl, H_2SO_4, HBr, and $HClO_4$) are "leveled" in aqueous solution by complete conversion to the hydronium ion, but they are distinguished as weak acids with strengths in the range of 1:9:30:160:400 in acetic acid (Sun, 2012).

7.6.4. Basic Solvents

The only thoroughly studied basic solvent is liquid ammonia, which has an extremely low ion product $[NH_4^+]$ $[NH_2^-]$ = 10^{-33}. As anticipated, this solvent has a significant neutralizing effect on acids. Due to the ion NH4$^+$, for instance, acetic, benzoic, nitric, and hydrochloric acids, all generate solutions with identical acidic properties, despite behaving very differently in water (McLuckey, 2013).

7.6.5. Aprotic Solvents

Purely aprotic solvents include hydrocarbons and their halogen compounds, as they do not react with additional acids or bases. The acid-base equilibrium in these solvents can only be studied with the addition of a second acid-base system that induces the normal reaction $A_1 + B_2 \rightleftarrows B_1 + A_2$. The vast majority of these experiments utilize an indicator as one of the reacting systems, even though the interaction between ions and molecules in these low-dielectric media makes interpretation challenging. Recently, the definition of aprotic has been widened to encompass solvents that can retain a proton despite possessing mildly basic properties. Some of these aprotic solvents, including N, N-dimethylformamide, dimethyl sulfoxide, and nitrobenzene, have high dielectric constants and can be used to dissolve a variety of substances. Their ability to distinguish between the characteristics of bases and acids is strong. It has been calculated that sodium methoxide dissolved in dimethyl sulfoxide produces a solution that is 109 times more basic than methanol because basic anions are faintly solvated in these solvents and function as extremely strong bases (Sun, 2012).

7.6.6. Concentrated Aqueous Acids

Strong acids such as hydrochloric, sulfuric, and perchloric (HCl, H_2SO_4 and $HClO_4$) react similarly to solutions of the ion H_3O^+ in water, and the acidity of these solutions increases with concentration. At concentrations greater than one molar (one mole of acid per liter of solution), the acidity, as measured by the effect on indicators or catalytic capacity, increases much more rapidly than the concentration. A 10^{-1} molar solution of any potent acid, for instance, is 1,000 times more acidic than a 1-molar solution. The hydronium ion H_3O^+, is known to have a strong propensity for further hydration, most likely to the $H_3O^+(H_2O)_3$ (that is, $H_9O_4^+$), and a decrease in the solution's water content increases its capacity to donate protons. This pattern is caused by is a measurement of the solvent's action on a basic indicator I, is used to determine the acidity of these concentrated solutions. It is defined as $H_0 + pK_{IH}^{+} - \log_{10} [IH^{+}]/[I]$ and corresponds to the pH of a diluted solution. It is frequently observed that the acidity function H_0 provides an estimate of the catalytic activity of an acid solution and is independent of the type of indicator. H_0 values for combinations of sulfuric acid and water with sulfuric acid concentrations ranging from 10% to 100% lie between 0.3 and 11.1, corresponding to an acidity range of approximately 11 powers of 10 (McLuckey, 2013).

7.6.7. Lewis Acids

Lewis acid-base equilibriums are considerably less well-known than conventional acid-base equilibriums, although the conditions for the former are more complex. When a particular Lewis acid interacts with several similar bases, the equilibrium constants adjust in tandem with the average basic strengths. This occurs, for instance, when a zinc halide, ZnX_2, interacts with a series of amines. In general, it is not possible to predict the degree to which a specific pair of Lewis acids and bases will react by arranging them in a particular order. Although the hydroxide ion (OH^-) is a much more powerful base than ammonia (NH_3) in proton acid interactions, the complex $Ag(NH_3)_2^+$ is reasonably stable, whereas AgOH is completely dissolved in reactions with the Lewis acid Ag+. However, the order is reversed for other metal cations (Sun, 2012).

As a result of this behavior, Lewis acids and bases have been classified as "hard" and "soft"; in general, hard acids preferentially react with hard bases, while soft acids preferentially react with soft bases. The terms were hard and soft are intended to convey that the atomic structures of rigid and impermeable hard acids and bases are distinct from those of more malleable soft acids and bases. Hard acids contain carbonium ions in addition to the ions proton, sodium, calcium, and aluminum. Soft acids consist of halogen cations, including cuprous, silver, and mercurous. Soft bases consist of iodide, thiocyanate, sulfide, and triphenylphosphine, while hard bases consist of hydroxide, fluoride, and numerous oxyanions. The theoretical justification for the fuzzy distinction between hard and soft categories is obscure. Based on the preferred reactions of hard acids with hard bases and soft acids with soft bases, a staggering quantity of empirical information can be coordinated (McLuckey, 2013).

7.7. THE EFFECT OF THE MOLECULAR STRUCTURE

The periodic table, a pattern formed when the elements are arranged in ascending atomic number order, is the most effective tool for identifying patterns and trends in element properties. When comparing the acidity of the hydrides of the various elements in the table, only those on the right-hand side, particularly the halogen elements (fluorine, chlorine, bromine, and iodine), manifest a significant degree of acidity.

This generalization is strengthened when the elements in the first portion of the periodic table are examined sequentially; as one approaches the right side of the table, carbon, nitrogen, oxygen, and fluorine are encountered. The acidity of the hydrides of these elements increases. Methane, CH_4, a carbon hydride, possesses no discernible acidic properties, and its pKa value is much more lower than those of ammonia (NH_3), water (H_2O), and hydrogen fluoride (HF), which have pKa values of 35, 16, and 4 respectively. The acidity of the hydrides in each descending group of the periodic table increases. In decreasing order, the elements oxygen, sulfur, selenium, and tellurium, along with fluorine, chlorine, bromine, and iodine, are included in the two categories on the right side of the table. Hydrogen telluride (H_2Te), hydrogen sulfide (H_2S), hydrogen selenide (H_2Se), and water (H_2O) have the following pKa values: hydrogen sulfide (H_2S), hydrogen selenide (H_2Se), and water (H_2O). Hydrogen fluoride (HF) is a mild acid, whereas hydrogen chloride (HCl), hydrogen bromide (HBr), and hydrogen iodide (HI) are potent acids in aqueous solution. Changes in bond strength, electronegativity (the ability of the atomic nucleus to attract electrons), and ionic solvation energy cause these tendencies, with the first being the most significant. When a hydride can lose two or more protons due to an increased negative charge on the base (Sun, 2012).

In a neutralization reaction, an acid and a base combine to generate salt and water. During the process of neutralization, H^+ and OH^- ions combine to form water. Acid-base neutralization is the most common type of neutralizing reaction.

7.8. APPLICATIONS OF NEUTRALIZATION REACTION

In a chemical titration, the neutralization point is used to identify unknown acid or base concentrations. Using a pH indicator or pH meter, we can precisely locate the site of neutralization. The molarity of the undetermined particles can be determined using simple stoichiometric calculations and the volume and molarity of the known material. To determine the unknown quantity of acids or bases, titration procedures are carried out. You can use a pH meter or a pH indicator with a distinct color change to determine the neutralization point. Simple stoichiometric calculations using the known volume of the unknown and the known volume and molarity of the extra chemical determine the unknown's molarity (McLuckey, 2013).

1. **Treatment of Wastewater:** The vast majority of industrial effluents contain a high level of toxicity, which is detrimental to the environment. Before discarding them, we must reduce their toxicity. Depending on the application, various chemicals are employed. Examples include sodium bicarbonate, magnesium oxide, calcium oxide, and calcium carbonate. Widespread use of chemical neutralization techniques in wastewater treatment reduces the potential damage an effluent may cause when released into the environment. Calcium carbonate, calcium oxide, magnesium hydroxide, and sodium bicarbonate are common substances used to control pH. Depending on the application, the ideal neutralizing chemical is chosen (Sun, 2012).

   The heat generated by a neutralization reaction is used to accelerate the chemical reduction of metal constituents during the synthesis of nanomaterials. In the chemical synthesis of nanomaterials, the neutralization reaction's heat can be used to enhance the chemical reduction of metal precursors.

   Before food can travel between the stomach and the intestines, our digestive systems must neutralize it. However, an alkaline environment is necessary for the absorption of nutrients through the intestinal membranes. This advantageous circumstance is the result of the development of antacid bicarbonate. There are numerous applications for neutralizing acidic and basic substances. One prevalent application involves antacid medications. These are intended to neutralize excessive gastric acid (HCl), which may be causing discomfort in the lower esophagus or stomach. This can also be remedied by orally ingesting sodium bicarbonate ($NaHCO_3$). In addition, sodium bicarbonate is frequently used to treat acid wounds and spillage in laboratories. When food travels from the stomach to the intestines, neutralizing reactions are also utilized. The pancreas produces antacid bicarbonate to provide the alkaline environment necessary for the absorption of nutrients through the intestinal membrane.

2. **Soil pH Management:** For optimal plant growth, every soil must meet specific parameters. Calcium Carbonate (Limestone) and Calcium Hydroxide (Slaked Lime) are two examples of compounds that can be added in acidic soil to neutralize it. Fertilizers and regulating the pH of the soil are two additional, albeit less well-known, applications. Incorporating hydrated lime (calcium hydroxide) or limestone (calcium carbonate) into soil that is too acidic for plant growth is possible. By neutralizing sulfuric acid (H_2SO_4) or nitric acid (HNO_3) with ammonia gas (NH_3), respectively, ammonium sulfate and ammonium nitrate are produced. These compounds are utilized in the production of fertilizer (McLuckey, 2013).

   When sulfur dioxide gas, a byproduct of coal combustion, reacts with water vapor in the air, sulfuric acid is formed and precipitates as acid rain. A scrubber is used to capture sulfur dioxide from smoke columns to prevent the emission of the gas. Initial calcium carbonate bombardment of the combustion chamber disintegrates into the lime and CO_2. This lime then reacts with the produced sulfur dioxide to form calcium sulfite. After removing the calcium sulfite and any unreacted sulfur dioxide, the mixture is treated with a slurry of lime suspension (McLuckey, 2013).

CHAPTER

8 Single Replacement Reactions

Contents

8.1. INTRODUCTION

The ability of an atom to participate in chemical reactions is due to the electrons that are in orbit around its nucleus. Chemicals of any kind can interact with one another, from atoms and ions to complex compounds. There are many different ways that chemical reactions can occur, and single replacement reactions are one such reaction. A reaction whereby one element is substituted for another in a compound is known as a single-displacement reaction or a single replacement reaction. Pure elements, like hydrogen gas, pure zinc metal or, and aqueous compounds, are key starting points. Products of a replacement reaction form a brand-new aqueous compound and pure element (Swenson, 2014).

A single-replacement (also known as a single-displacement) reaction generally takes the following form: A+BC→AC+B. In this overall reaction, element A (metal) takes the place of element B (metal) in the compound, which is also a metal. When the element doing the replacement is a nonmetal, it should replace another nonmetal in the compound. Y+XZ→XY+Z (where Y is a nonmetal replacing the Nonmetal-Z in the compound with X). When a less reactive element within the compound can be replaced by a highly reactive element, it's predicted that a single replacement reaction will occur. The reactivity series ranks elements according to how reactive they are to reactions, such as single-replacement reactions. In reactivity series, the more reactive elements substitute the less reactive ones, not the opposite one. There are isolated rankings for elements forming cations and anions. The reactivity series for cation-forming elements is more extensive, while the trends are not quite direct (Hyeon, 2013) (Figure 8.1).

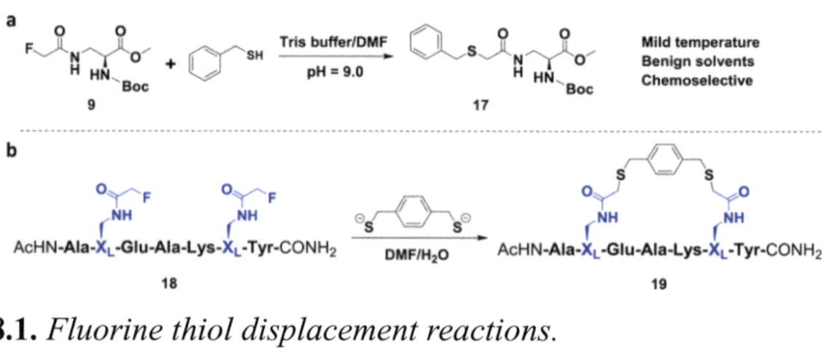

Figure 8.1. *Fluorine thiol displacement reactions.*

Source: Islam, et al., (2022). Unprotected peptide macrocyclization and stapling via a fluorine-thiol displacement reaction. Nature Communications. – Scientific Figure on ResearchGate. Available from: https://www.researchgate.net/figure/Fluorine-thiol-displacement-reactions-FTDRs-a-A-model-reaction-between-benzyl-thiol-and_fig1_357898933.

Reactivity is really complex phenomena. Since there are so many different forms of reactions, they should really be ranked in order of importance. The reactivity series considers an element's reactivity with acids and water, including the ease where which it loses electrons to produce cations (Swenson, 2014).

8.1.1. Other Definitions

8.1.1.1. Chemical Reactions

Chemical reactions are the basis of all life processes, including changes in the nonliving parts of various environments across the planet. Chemical species, whether molecules, atoms, or complex compounds, interact with each other and transform into various chemical species during chemical reactions. While some reactions can occur without the use of energy, others require the removal of an energy barrier before they can proceed.

8.1.1.2. Reaction Types

There are various ways by which chemical species can mix with each other when reactions occur. A new chemical compound is created when two or more chemicals react in synthetic reactions. In contrast, a more complex compound actually disintegrates into two or more basic substances during decomposition. Both single-and-double replacement reactivity results in the transformation of the original reacting compounds into new product compounds through the exchange of chemical species between the reacting substances (Hyeon, 2013).

8.2. SINGLE-REPLACEMENT REACTION PRODUCTS

1. What two elements could be substituted in proposed reactions?

   In general, Anions and cations can be substituted for one another in a compound by elements that form them. To determine what type of ions an element might develop, the following guidelines should be considered.

- Metals typically produce cations. This comprises elements from groups 1 and 2, some elements from groups 13 and 14, and transition metals.
- The group 17 elements, which typically form 1^- charged anions, are the most prevalent non-metals in single replacement reactions.

- The cation H^+ is typically produced by hydrogen in a single replacement reaction.

  Because copper metal is a transition metal, it is likely that the reaction between aqueous silver (I) nitrate and copper metal will result in the formation of copper cations. In the compound $AgNO_3$ (aq), the silver cations can be replaced by the copper cations to create a new compound (Swenson, 2014).

2. Which new compound will result in a product?

   We can predict the products that can be formed once we know which element in the ionic compound can be substituted. Cu $(NO_3)_2$(aq) can be made by replacing the silver atoms in $AgNO_3$ (aq) with copper. Ag(s), elemental silver, and other byproducts would also emerge from the process. The complete and balanced reaction can be written as follows:

$$2AgNO_3\,(aq) + Cu(s) \rightarrow Cu\,(NO_3)_2(aq) + 2Ag(s)$$

8.3. DETERMINING THE PROBABILITY OF SINGLE REPLACEMENT REACTION

Understanding the relative reactivities of the two elements C & A in the generic sequence above, or copper and silver in the reaction—we can predict whether a single replacement reaction will happen once we know which elements might be substituted in the single displacement reaction. There will be no reaction if element C is highly reactive compared to element A.

The reactivity series, known as the activity series, categorizes elements according to their reactivity for specific kinds of reactions, including single-replacement reactivity. The more reactive elements can replace the less-reactive elements in the reactivity series. Different rankings for elements comprising cations and anions exist (Hyeon, 2013).

The following is the order of reactivity, from most reactive to least reactive, for elements that are more likely to acquire electrons in order to develop anions:

Most reactive $F_2 > Cl_2 > Br_2 > I_2$ to Least reactive

For these elements, group 17 of the periodic table can be considered. The greater the element's ranking in the column, the highly reactive it would be. The reactivity series for the cation-forming elements is longer, while the trends aren't as straightforward (Swenson, 2014).

8.4. TYPES OF SINGLE-REPLACEMENT REACTIONS

8.4.1. Cation Replacement

Cations are positively charged ions or metals. Potassium, calcium, magnesium, and sodium are four key cations present in the body. In this form of reaction, one cation substitutes another. Cation exchange is the process whereby cations are adsorbed from solutions in a reversible manner on the charged surfaces of sediments. The phyllosilicates' isomorphous replacement and broken edges, in addition to the organic matter's deprotonation of acid groups, produce a net negative charge. The electrostatic attraction of cations helps to cancel out the negative charge. Cation exchange is a quick and flexible process, and cations are substituted on equivalent charge principle. Cation exchange is influenced in some way by every sediment component. But the most important aspects of this process are organic content and minerals in the clay portion (<2 μm). The alkaline earth and metal cations (Mg^{2+}, K^+, Na^+, and Ca^{2+}) do not form strong secondary minerals and remain in the solution phase to balance the sediments surface charge. This is due to their charge and size. Al^{3+}, Al monomers, and H^+ are also found on the exchange sites when the environment is acidic (Hyeon, 2013) (Figure 8.2).

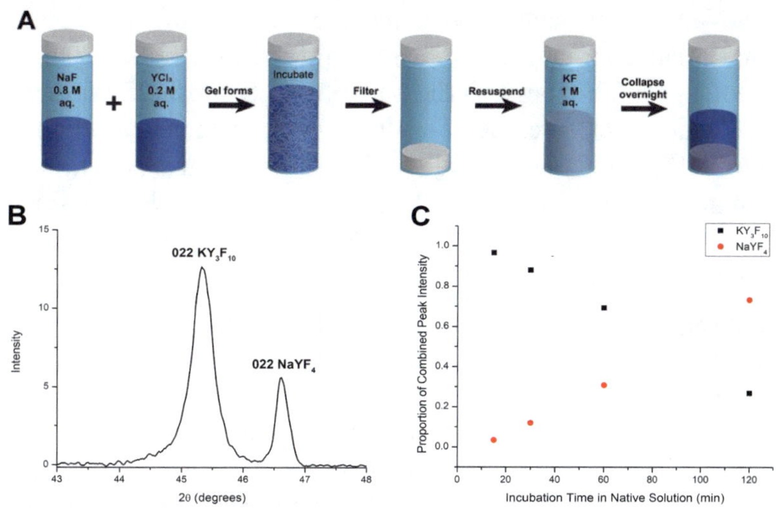

Figure 8.2. *Ion replacement experiment.*

Source: Pauzauskie, et al., (2021). Multi-Step Crystallization and Chemical Evolution of Sodium Yttrium Fluoride. – Scientific Figure on ResearchGate. Available from: https://www.researchgate.net/figure/Ion-replacement-experiment-A-Schematic-for-the-cation-replacement-experiment-B_fig1_356774238.

Ø ***Examples:***

- Hydrochloric acid (HCl) can safely dissolve zinc (Zn), releasing hydrogen (H_2) gas.
  Zn (s) and 2 HCl (aq.) → $ZnCl_2$ (s) + H_2 (g)
- Copper (Cu) can replace silver (Ag) in aqueous solutions of silver nitrate ($AgNO_3$).
  Cu (s) + 2 $AgNO_3$ (aq.) → Cu $(NO_3)_2$ (aq.) + 2 Ag (s/ppt.)

8.4.1.1. Cation-Exchange by In-Situ Electron Microscopy

Transmission electron microscopy (TEM), also in scanning mode (STEM), and its ancillary analytic methods, Energy Dispersive X-Ray Spectroscopy (EDS) and Electron Energy Loss Spectroscopy (EELS), have all undergone significant advancements over the past two decades that have revolutionized the most widely used method for imaging and studying colloidal synthesis-produced nanostructures. Actually, TEM/STEM is currently able to image nanoparticles (NPs) with a point resolution of more than 0.1 nm, revealing their scope, morphology, crystal structure, crystal defects, crystal strain, and composition even at the scale of a single atomic column. A number of technical advancements have contributed to the development and availability of (S)TEM enhanced imaging and analytical functions: spherical aberration (CS) correctors for TEM and STEM, novel EELS and EDS spectrometers, and novel CMOS electron cameras—all of which can detect a single electron with sub-pixel precision—provide either ultrasensitive or rapid image acquisition and storage functions. The massive impulse also obtained by its in situ functions over the past few years was the natural consequence of the (STEM) field's constant and fast evolution.

In essence, performing in situ (S)TEM entails subjecting the sample to one or multiple external stimuli during its observation and studying the equivalent changes with appropriate temporary resolution. Today, a variety of stimuli, comprising heating in controlled gaseous atmosphere, electrical biasing, mechanical indenting, cooling or stretching, liquid immersion, illumination, and electrical biasing in electrochemical cells, can be applied to (S)TEM samples while inside the microscope (Swenson, 2014).

With this respect, an in-situ heating experiment needs a committed sample holder suitable for expanding, controlling, and in the long-term maintaining the sample temperature in a controlled way for extended timeframes. Two essential technical prerequisites must accompany this capacity: reducing the sample's thermally produced spatial drift and the power needed to heat it. Over the last 10 years, these heating holders were either made to order or purchased. The heat source was a small furnace that was surrounded by a clamped TEM sample. The sample was particularly deposited on an electron-transparent film and supported by a metallic, thermally conductive grid, or directly attached to the grid (Hyeon, 2013).

A thermocouple in contact with the holder's heating element or an *a prior* calibration curve based on the relationship between temperature and heating current can be used to measure temperature. The heating tip may be designed to be thermally isolated from other sections of the holder rod, thus limiting heat conduction and dissipation, and these holders were built to minimize the furnace and sample's thermal expansion. However, they unavoidably had three main drawbacks. First, the furnace size and the metal grid that support the sample mean that their thermal dilatation and TEM sample's drastic spatial drift could not be prevented during thermal ramps. This eventually leads to the thin film that supports the sample breaking locally. Because of the magnitude of drift effect, sample imaging as the temperature changes can be difficult, and simultaneous EDS or EELS compositional inquiry and mapping was essentially impossible. Second, the temperature can be measured on the heating element rather than the heated sample, resulting in an unavoidable source of uncertainty regarding the actual thermal value the image sample effectively reached and measurement errors up to a few tens of degrees. Third, the enormous amount of infrared (IR) radiation formed by the hot sections saturated the EDS detector, rendering any EDS-based analysis impossible once the heating furnace and sample reached a steady temperature of over 80–100°C and appropriately low spatial drift to allow imaging and EELS-based analysis or mapping.

In contrast, the furnace-based model for thermal annealing has been completely abandoned in favor of MEMS (micro electro mechanical systems) chips technology in recent generations of in-situ sample holders, which have been developed and made commercially accessible. The electron-transparent thin film that served as the sample substrate in the traditional metal grid fastened in the middle of the heating furnace has been uncontrolled in these sample holders in favor of a multipurpose chip that combines the sample support role with the heating and local temperature measuring functions. Most MEMS are made of thermally conductive materials that can reach temperatures greater than 1000°C. They also have one or multiple electron-transparent zones that can be left empty, covered with a carefully chosen material like amorphous carbon, for example, or made of thinner parts of the same chip serving as a supporting film for the sample. Two small concentric spiral wirings encircling the electron-transparent zones on top of the MEMS' main body typically serve as the heating and temperature measurement elements. They are linked to an external controller that continuously measures the current local temperature, and provides the driving current necessary to achieve the desired temperature. Compared to the previous holders that were based on furnaces, this architecture offers several advantages: Because the heating element is a real component of the heating tip comprising of the entire MEMS, they dilate and contract together. This keeps the x-y spatial drift during the thermal ramps to a few nanometers and the z drift, which is limited by the MEMS's bulging and unbudging during its heating and cooling ramps, to a minimum (Swenson, 2014).

In addition, the furnace-based design's inherent uncertainties in monitoring the sample's actual temperature were mitigated by the heating element's small size, relation with the temperature measuring spiral, and correspondence to the sample deposition zone. Furthermore, the heating elements' small size provides two additional benefits that both of which are based on the extremely low power required to increase or decrease the formed heat and the corresponding thermal variation: Thermal ramps can be completed in a few tenths of a second, and the heating MEMS produce much less IR radiation than furnace-based in situ sample holders do. This makes it possible to conduct EDS-based analyses even at high temperatures and, where possible, at a specific tilting angle so as to create 3D chemical maps.

Meanwhile, colloidal synthesis has developed into a simple and profitable method for the production of a wide variety of nanomaterials, allowing for unprecedented control levels over their chemical and physical properties and the adjustment of their size, composition, shape, and related crystal structure. In this perspective, (S)TEM arose as the most important imaging and analytical instrument for comprehensively understanding the relationship between these nanostructures' design, shape, and chemical properties. Furthermore, the fine characterization of colloidal NPs was greatly improved by advances in (S)TEM, though a few limitations remained. In fact, because colloidal NPs are formed in a liquid environment by single or multi-step reactions, the only method to determine what was happening during their growth with (S)TEM was taking a sufficient ratio of aliquots from the running reaction(s) and examine ex situ the changes in their intermediate products, effectively "freezing" the reaction(s) at various intermediate steps prior to the production of the final nanomaterial(s) (Hyeon, 2013).

Otherwise, only the intermediate products and finished product could be examined in-depth if the reaction started at least partially with solid precursors as reagents, or started slowly enough to produce solid intermediate products. To put it in a simpler way: In very fast liquid reactions, intermediate steps must be lost. The fast, direct reaction of a starting population of ionic NPs with one or multiple cationic species dispersed in a common liquid solution is normally the method used to obtain ionic NPs that are tougher to produce through direct synthesis.

8.4.2. Anion Replacement

Primary anions include chloride, bicarbonate, proteins, and phosphate. Anions are negatively charged ions or nonmetals. One anion substitutes another in this type of reaction. Nevertheless, anion exchange is unproductive on positive ion charges such as iron and magnesium (Swenson, 2014) (Figure 8.3).

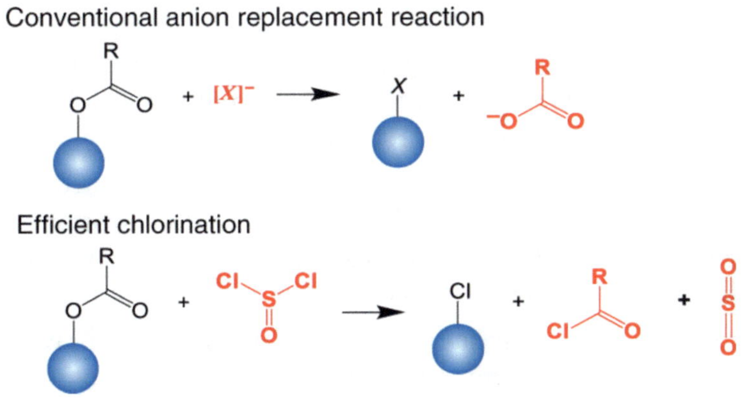

Figure 8.3. *Naturally occurring anion mineral stone.*

Source: Li, et al., (2018). Bright colloidal quantum dot light-emitting diodes enabled by efficient chlorination. Nature Photonics. – Scientific Figure on ResearchGate. Available from: https://www.researchgate.net/figure/Chlorination-ligand-exchange-mechanism-Chemical-equations-for-conventional-anion_fig1_323399838.

Ø *Examples:* A sodium bromide (NaBr) compound will have chlorine (Cl_2) replace bromine (Br_2) because chlorine is more reactive compared to bromine.

$$Cl_2\ (aq.) + 2\ NaBr\ (aq.) \rightarrow 2\ NaCl\ (aq.) + Br_2\ (aq.)$$

The process by which acidic anions are absorbed by basic substances is known as anion exchange. It defines the ion exchange (since chloride or hydroxide) for one or multiple anions (such as sulfate) through the exchange of ions. It is greatly effective on negatively charged particles, for example, Bicarbonate.

Ion exchange is a method of chemistry separation that allows for the selective retention of different ionic substances on ions for resin exchange. In this respect, resins are water-insoluble polymeric ionic materials whose ions can be exchanged in locations on the polymer-chain known as exchange sites.

Water goes through these exchange sites along a device that helps ionization. This results in the conversion of unwanted ions into desirable ones. Exchange units are divided into two categories: cations and anion groups. In the last option, resin beads are typically saturated with sodium or potassium. In anion exchange, beads are saturated across negatively charged ions in the exchange of contaminants.

An excellent tool for exchanging or eliminating contaminants in low concentrations, ion exchange can assist in determining which ion exchange unit is most suitable for treatment (Hyeon, 2013).

Calcium removal, which is the primary action associated with scale formation, is the most common method for controlling corrosion and scale. This process inhibits the probability for scale to frame. Calcium ion removal is normally done using a base-exchange water conditioner.

Ion exchange resins in base-exchange water softeners can, when mixed with water, transform calcium ions into sodium ions. It is fully automated thanks to highly dependable control valves that automatically regenerate the softener based on the operating time or treated water capacity.

8.5. DIFFERENCES BETWEEN ANIONIC AND CATIONIC SINGLE REPLACEMENT

Anions are atoms with a final negative charge. This indicates that the atom has gained one or multiple charged electrons from a secondary atom or molecule, and thus now carries an excessive negative charge. Cationic single replacement anions are the opposite. They have a positive charge because they lost one or multiple electrons and the protons' positive charge within the nucleus is not balanced. Through ionic bonding, cationic and anionic species can draw one another and create a new molecule (Swenson, 2014).

8.5.1. Anionic and Cationic Single Replacement

An anion reacts with a different ionic molecule in anionic replacement. As the reaction progresses, the ionic molecule, which is made up of an anion and a cation, loses its anion and is replaced by the new reacting anion. In cationic replacement, cations react with ionic particles comprising of anions and cations and, additionally, a switch occurs, with the new cation substituting the old cation. The new ionic molecule and the replaced species are released in both instances.

8.6. GROUPS OF SINGLE-SUBSTITUTION REACTIONS

8.6.1. Metal Replacement

If the metal in a single response reaction is highly reactive compared to the metal in the substance, the metal in the substance will be exchanged. The list of metallic elements in the metal reactivity series is organized vertically, with the most reactive metals at the top. Moving across the reactivity series column, the metal's relative reactivity will reduce. Thus, in the reaction metal will be replaced if the reacted metal is greater on the reactivity series. This indicates that the metal will form a bond with the compound's negative ion on its own. Consider two examples (Hyeon, 2013).

Ø *Example 1:*

$K + Na_2 O \rightarrow$

Potassium is higher on metallic reactivity series than sodium, and that implies potassium will replace sodium. The reaction formula is:

$2K + Na_2O \rightarrow 2Na + K_2O$

This sub-class of single-replacement reactions is referred to as a metal replacement reaction since a metal (sodium) is being replaced.

8.6.1.1. The Theory of Cementation (Metal Displacement) Reactions

Ions are reduced to zero-valent at a solid metallic surface during cementation, a precipitation process. According to Power & Ritchie (1976), the metallic surfaces serve as an oxidizing agent and supply the electrons needed to reduce copper from solutions. The cementation process is frequently used to recover copper from electrowinning, electroplating, and ore leaching solutions. In the process of purifying and recovering metallic ions from process solutions or industrial waste streams, cementation reactions are crucial.

At a solid-metal interface, cementation is heterogeneous whereby ions are reduced to zero-valence. As an ion exchange, it is an oxido-reduction reaction that dissolves the sacrificial metal and cements the ionic mineral, transforming it into a metallic powder (Swenson, 2014).

Using copper as a reductor metal and iron as sacrificial metal, cementation produces the following physical reaction:

$Cu_2^+ + Fe \rightarrow Cu + Fe_2^+$

Cementation, in which the precipitated solid copper metal was recovered as flakes or powder on the scrap iron surface, was once a useful process for the production of copper. Although this hydro-metallurgy primary process has been replaced by electro-wining, cementation is still extensively used industrially to recover different heavy metals, and the gold cementation by zinc in the (Merrill Crowe) process, which covers for a significant ratio of the globe's gold production.

8.6.1.2. Origins

The method was originally used in Bavaria in 1601 and probably began in Bohemia during the 16th century. Mathias Meysey and William Ellyot patented the method in England in 1614. At that moment, the "invention" may just be the launch of a new business, new product, or even a monopoly. They later offered the patent to Sir Basil Brooke, but in 1619 he was made to give it up. Because he was unable to supply the required quantity of high-quality steel, it was determined that a patent clause that forbade the steel import was undesirable.

The Brooke furnaces were probably located in his Madeley at Coalbrookdale manor, which was in existence prior to the English Civil War and was recently excavated. Bar iron probably came from the Forest of Dean, where he and a partner once worked as partners in farming the King's ironworks. By 1631, it was known that Swedish iron was among the best raw materials. This was also true for some brands, like the double-bullet from Osterby. The brand was owned by Louis De Geer and his descendants. These were among the primary ironwork investors in Sweden to utilize the Walloon procedure of fining iron, creating what was referred to in Britain as foreground iron. It got its name from the Regrund Swedish port, north of Stockholm, where most of the ironworks were located in the hinterland. The Dannemora mine ultimately supplied the used ore (Swenson, 2014).

8.6.1.3. Process

Wrought iron and charcoal serve as the foundation for the process. In a furnace, it makes use of one or more long stone pots. Each was typically 14 feet by 4 feet and 3.5 feet deep in Sheffield. Iron bars and charcoal are pressed in rotating layers, with the top level of charcoal and afterward refractory matter to produce a pot air tight container. Cement powder, a mixture of powdered charcoal, mineral salts, and soot used by some producers, gave the process its name. Each cycle can treat up to 16-tons of iron in larger works (Hyeon, 2013).

The pots were afterwards heated from below for 1 week or more, depending on the iron bars thickness. Bars were examined on a regular basis, and when they were in good condition, the heat was turned off and the pots left to cool for about 14 days. The iron had "added" about 1% of its mass due to the charcoal's carbon, resulting in a variety of blister steel bars.

Shear steel was produced by shortening, binding, heating, hammering, pressing, or rolling the bars. They could also be disintegrated and melted in a crucible with a flux in a crucible furnace to make crucible steel or cast steel, a technique Benjamin Huntsman invented in the 1740s.

8.6.2. Hydrogen Replacement

Hydrogen gas is one of the byproducts of many metals' reactions with acids. Hydrogen gas and zinc chloride are the results of a rapid reaction between zinc and hydrochloric acid. Zinc metal displaces hydrogen in a single replacement reaction to produce zinc chloride and hydrogen gas. The reaction is exothermic so, creates immense heat: $Zn\ (s) + 2HCl\ (aq) \rightarrow ZnCl_2\ (aq) + H_2\ (g)$.

An active metal takes the place of an acid's hydrogen in hydrogen-replacement reaction. Some metals are highly reactive such that they can take the place of water's hydrogen. The results of such reactions are hydrogen gas and the metal hydroxide. All group 1 metals go through hydrogen replacement reaction. According to the equation, sodium and water react to produce sodium hydroxide and hydrogen gas: $2Na(s) + 2H_2O(1) \rightarrow 2NaOH(aq) + H_2(g)$

8.6.2.1. Hydrogen Metallurgy

Hydrogen metallurgy technology uses hydrogen rather than carbon as a reduction-agent to decrease CO_2 emissions. Hydrogen use is beneficial to the promotion of the steel business's sustainable development. Different applications for hydrogen metallurgy exist, including H_2 reduction iron production in Japan, hydrogen-based and ULCORED steelmaking in Europe; and the United States' hydrogen flash ironmaking technology (Hyeon, 2013).

During the BF process, a fraction of the coke used in hydrogen reduction ironmaking is replaced with hydrogen to cut CO_2 emission. Carbon emissions were down 9.4% in the first stage of experimental BF operations with H_2 injection from 2014 to 2016 compared to the non-H_2 injection stage. Due to the low density of H_2 and the endothermic reactions that accompany hydrogen reduction, the adjustable position of BF in the stack and channel as well as the preheating of H_2 prior to injection are carried out to guarantee maximum reduction performance as well as a stable internal temperature. In the second stage, a more extensive test is performed to simulate a real-world BF of 4000–5000 m^3 progressively. By 2030, it is estimated that the first BF will use H_2 reduction, and by 2050, the technology will exist in Japan. In addition, the production of hydrogen from COG has been the focus of the COURSE50 project. The temperature of COG when it leaves a carbonization chamber can reach 800°C, maximizing the application of sensible heat to catalyze the cracking of tar and hydrocarbon compounds and the subsequent production of H_2. The gas volume doubles and the H_2 content of COG rises from 55 vol.% to 63 vol.%–67 vol.% as a result of this modification (Swenson, 2014).

COURSE50 is the original step, according to the report from the 12th CSM Steel Congress by Inoue, Vice President (VP) of Nippon Steel Co. A number of relevant hydrogen metallurgy technologies, for instance H_2 reduction in BF (internal H_2, external H_2), H_2 reduction without BF, CO_2 capture and storage (CCS), and CO_2 capture utilization (CCU), these have been done or are currently being planned in Japan.

8.6.2.2. Hydrogen-Based Steelmaking

This method uses H_2 as a reducing agent to make hydrogen iron. In the production of steel, hydrogen can be utilized in one of two ways: as the sole reducing agent in the direct reduction of iron (DRI) process (H_2-DRI) or as a secondary reducing agent in the BF-BOF route (H_2-BF).

Electrolysis of water produces H_2, and water constitutes the tail gas. The fossil fuel byproduct of the shaft heater is insignificant; nevertheless, when electric power is taken into account, the entire process's CO_2 emissions per ton of crude steel only comes to 300 kilograms, which is an 84% decrease when compared to the 1850 kilograms produced by a traditional BF process.

8.6.2.3. Steel Production Through the BF-BOF Process

The BF-BOF route, otherwise called the primary production route, represents 60% of steel manufacturing in Europe. The blast furnace and coke plant produce most of the emissions. Coking coal is produced at the coke plant and is applied in the blast furnace to reduce iron and provide heat. Because it only produces water rather than carbon-dioxide after reacting with iron ore, H_2-BF has the potential to cut emissions in both the blast furnace and coke plant. Pulverized coal (PC), natural gas, oil, or a combination of both, altogether produce CO_2, and are currently the most popular auxiliary reducing agents. Nevertheless, because hydrogen cannot be used alone in a blast furnace for technical reasons, H_2-BF is frequently regarded as a transition to H_2-DRI for immediate emission reductions. H_2-BF will be utilized by a number of European steel producers (Swenson, 2014) (Figure 8.4).

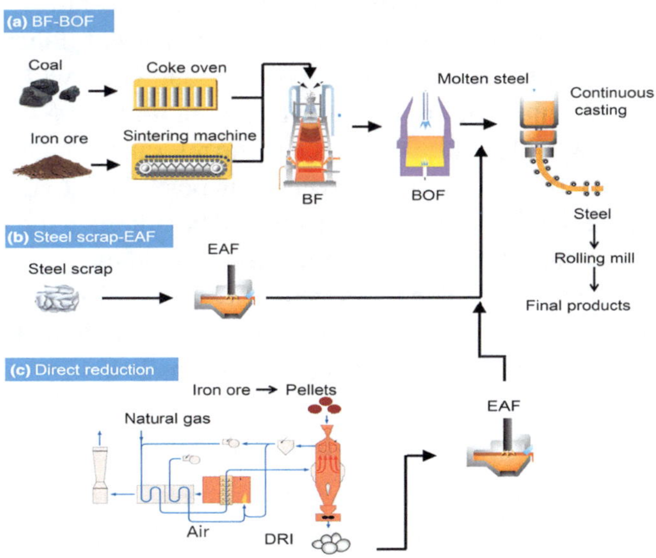

Figure 8.4. *Steel production using integrated BF-BOF process.*

Source: Jorge, P., Begoña, P., Manuel, B., Valerie, E., Pravin, K., Abhijeet, R., Pilar, L., & Luis, M. R., (2023). Integration of carbon capture technologies in blast furnace based steel making: A comprehensive and systematic review, Fuel, 336, https://www.sciencedirect.com/science/article/pii/S0016236122038984.

8.6.2.4. Hydrogen From Electrolysis

Of the various methods for producing hydrogen, the one that uses renewable energy and electrolysis must be given top priority because it reduces emissions by more than 21% when hydrogen is used in the BF-BOF. Only three of the eight projects were found clearly state their intention to use green hydrogen: Voestalpine in Austria (project H_2FUTURE), TATA in the Netherlands (H_2ERMES project) and ArcelorMittal in France, are involved. Thyssenkrupp anticipates having some green hydrogen by the mid-2020s; however, this will be contingent on how the ratio of renewable energy in the electricity grid develops. Emissions could rise up to 36.7% if hydrogen is created through electrolysis using electricity from the German grid.

8.6.2.5. Hydrogen Produced by Reforming Natural Gas (Grey Hydrogen)

Several organizations claim that they will apply green hydrogen when it is available in sufficient quantities and at reasonable rates. The date at which these conditions will be fulfilled under the current policy structures is unknown, but it can be schedule anytime soon. They will utilize grey hydrogen in the interim. For instance, the Duisburg project of Thyssenkrupp announced that it would acquire hydrogen in the interim through the regional hydrogen network of Air Liquide. Given their portfolio of H_2-production, the hydrogen is probably gray. Emissions would be minimized by only 2.1% using grey hydrogen, 10 times less compared to green hydrogen. Including CCS to the reform system (blue hydrogen) may lessen discharge levels like green hydrogen (for certain residual discharges). It is likely that this will be required to accelerate hydrogen production (Hyeon, 2013).

8.6.2.6. What Is Reduction Based on Hydrogen?

The chemical removal (reduction) of oxygen from iron-ore in its solid state is known as direct iron reduction. Currently, fossil fuels like natural gas or coal are used to chemically reduce iron from iron ore for use in the steelmaking process. Direct Reduced Ironmaking (DRI) is the scientific name of this method. Carbon consolidates with oxygen within the iron ore, creating metallic iron and carbon-rich process gas, as per the following basic chemical reaction: $2Fe_2O_3 + 3C = 4Fe + 3CO$.

Hydrogen can also be used to reduce iron ore rather than carbon; As a result of these reactions, water is the waste gas generated in this situation: $Fe_2O_3 + 3H_2 \rightarrow 2Fe + 3H_2O$, $FeO + H_2 \rightarrow Fe + H_2O$. In petroleum gas based

DRI manufacturing, hydrogen assumes a part in the reduction procedure, however this is in mixture with carbon. Gas-based DRI production has lesser greenhouse gas (GHG) emissions than the BF route, with 1.5 tons of CO_2 being released for every tons of DRI produced (Swenson, 2014).

8.6.2.7. Hydrogen Replacement Uses

Hydrogen is currently produced on a commercial basis. It is applied as a feedstock in the chemical sector and refineries, being part of a mixture of gases in the production of steel, and in the generation of heat and power. There are approximately 75 MtH_2/yr of pure hydrogen and 45 MtH_2/yr of mixed gases produced globally. This is comparable to Germany's annual energy consumption and amounts to 3% of worldwide final energy demand. Hydrogen is not an energy source but rather a versatile energy carrier. It can be made from a variety of feedstocks and used in almost any application.

Electrolysis can convert renewable electricity into hydrogen, which can be used to combine renewable energy with end uses that are harder to electrify. In smart electrification, electrolyzers can also complement alternatives like batteries, vehicle-to-grid, and demand response by providing flexibility to the grid using this coupling. It was estimated that by 2021, roughly 47% of the world's hydrogen production will come from natural gas, 27% from coal, 22% from oil (as a byproduct), and just 4% will come from electrolysis. In 2021, renewable energy constituted about 33% of electricity consumption globally, meaning renewable energy only constitutes 1% of hydrogen production worldwide. In 2021, the production of electrolytic hydrogen from dedicated production remained restricted to demonstrate projects with a capacity of 0.7 GW. In contrast, the 1.5°C Scenario would require 4–5 TW by 2050, which is faster than what solar photovoltaic (PV) and wind have achieved thus far. The use of hydrogen as an energy carrier remains limited, primarily to road vehicles (Hyeon, 2013).

By June 2021, around 40000 fuel cell electric vehicles were available for use across the planet, with nearly 90% of those found in four nations: Japan, the People's Republic of China, Korea, and the United States. There were be over 3100 fuel cell e-trucks and approximately 6000 fuel cell e-buses by the end of 2020, with 95% based of China. The global vehicle fleet is much larger than these sums. However, hydrogen's full value is only realized when it is further transformed into derivatives. Hydrocarbons and nearly any atom can be made by combining hydrogen and carbon from CO_2. It can be used to produce ammonia, which is often used as fuel for new applications such as shipping or fertilizer feedstock (for most current uses). Methanol, synthetic fuels, and even a reducing agent for coal replacement in iron production all make use of this technology (Swenson, 2014).

The increased energy density that results from its transformation into these products further reduces the cost of long-haul transportation and storage. As a result, global energy trade is effectively made possible by switching to hydrogen derivatives. Liquid ammonia, for instance, has a gravimetric energy density (MJ/kg) that is more than 20 times that of lithium-ion batteries and nearly eight times that of liquid ammonia's energy density (MJ/m^3). The increased energy density of hydrogen-derived goods effectively extends the distance over which energy can be transported economically, connecting affordable renewable energy aspects with demand centers with limited or expensive renewable energy potential.

Importing nations would be able to access cheaper (than domestic) resources through global energy trade using hydrogen derivatives, which would boost the system's resilience and energy security by providing alternative means of fulfilling final demand. Nevertheless, the financial gain will not be the only factor that will determine hydrogen trading. In the long haul, when technologies reach complete maturity and are dispatched at large, it is projected that importing nations will actually want to depend on different choices with a less cost range. As a result, non-economic factors will largely define trading partners (Hyeon, 2013).

8.6.2.8. Challenges

1. **Scale Up:** According to the IEA's core Sustainable Development Scenario (SDS), electrolytic hydrogen can be used as a primary reducing agent on a commercial scale and will reach 12 Mt by 2050. In addition, the IEA anticipates that India and China will have the greatest demand for electrolytic hydrogen in steel by 2050 (roughly 4.5 Mt of hydrogen in each) as a result of their large production volumes and access to a high quantity of affordable renewable electricity.

   Today, roughly 70 Mt of dedicated hydrogen are generated, with 76% coming from natural gas and 23% from coal. Water electrolysis comprises less than 0.1% of the world's dedicated hydrogen production. The annual demand for electricity would be 3,600 TWh, which is more than the annual production of electricity by the European Union (EU), if all dedicated hydrogen production were done through water electrolysis (using electricity and water to make hydrogen) (Swenson, 2014).

Global demand for hydrogen is expected to rise to 287 Mt by 2050 under the IEA's SDS, which is an increase of more than 400% from 2020. This presents a significant challenge for scaling up.

2. **Costs:** Renewable hydrogen is two to three times more expensive to produce than fossil fuel due to long-term average costs of USD 75/bbl for oil and USD 4–6/GJ for natural gas. Hydrogen pipelines can cost anywhere from 10 to 50% more. For road transportation, fuel cells and storage tanks cost multiple times more than internal combustion engines. Currently, aviation fuels made of synthetic materials can cost three to six times as much as jet fuel made from fossil oil. Ammonia can cost anywhere from 50 to 75% more than methanol, while steel can cost anywhere from 30 to 40% more, and methanol can cost 150% more than steel.

   The IEA discovered that innovative process routes such as CCS on the BF, gas based-DRI and smelt reduction, can be required to cost 10–50% more compared to commercially available counterparts in a particular regional context. This cost increase considerably exceeds steelmaking's current profit margins. The IEA analysis also found that the rate of generating hydrogen from renewable electricity may drop by 30% by 2030 due to the scaling up of hydrogen production and falling costs of renewable energy.

3. **Safety Concerns:** When applied on a large scale, hydrogen poses the same health and safety risks as other energy carriers. Hydrogen handling needs specialized equipment and procedures because it is a light gas and small molecules. Because hydrogen is especially small, it can get into some materials, such as some forms of iron and steel pipes, and make them more likely to break. Additionally, it is less likely to escape through connectors and sealings than bigger molecules like natural gas (Hyeon, 2013).

   Steel vessels and pipes can also become brittle and crack from hydrogen as they get reduced. Austenitic tempered steels don't experience the negative effects of hydrogen embrittlement. Process Safety Management systems are already being developed and implemented by steelmakers to manage the risk of toxic or flammable hazardous material loss from containment. When hydrogen is used, risks associated with it must be incorporated into risk assessments and associated controls.

4. **Infrastructure:** There are only approximately 4500 kilometers of hydrogen pipelines worldwide. As a light and atomically small gas, hydrogen can be tough to contain, and specialized infrastructure may have to be developed to allow for scale distribution, using renewable resources from remote locations would need additional investment in transport infrastructure, comprising of pipelines, conversion, and liquefaction units, and storage. Current high-pressure natural gas transmission pipes may be converted to provide pure hydrogen in the future when no longer used for natural gas. However, their suitability must be assessed on a case-by-case basis and will depend on the steel grade used in the pipeline and the purity of hydrogen being transported. Another setback is that three times more capacity is required to supply the same quantity of energy as natural gas. Electrolysis requires sufficient electricity and water. About 1 kilogram of H_2 requires roughly 9 liters of water, which also generates 8 kilograms of oxygen as a coproduct. In areas where there is a lot of water, this might be a challenge (Swenson, 2014).

5. **Policy:** The road transportation sector, especially fuel cell e-vehicles and refueling stations, has received major policy discussions till date. Policy attention is changing to comprehensive national strategies, infrastructure, hydrogen supply, and industry adoption as the focus shifts to a broader range of hydrogen applications. Additionally, there is no established method for distinguishing low-carbon hydrogen from hydrogen derived from fossil fuels. As a result, consumers have no way of knowing where the hydrogen that is produced comes from or how it affects the environment since there is no link between incentives (market and policy) and production. This gap could be filled by a number of modern hydrogen certification efforts.

8.6.3. Halogen Replacement

Ø **Group 1: The Alkali Metals:** Every alkali metal vigorously reacts with halogens to generate salts, the most important of which are NaCl and KCl for industrial use. Sodium chloride is used to melt ice on roads (via freezing point depression) and as a meat preservative. Because potassium has a positive effect on plant growth, KCl is essential for plant fertilizers. All of these metal halides, with the exception of LiF, which has a high lattice enthalpy due to the high electrostatic attraction between Li^+ and F^- ions, are white ionic crystalline solids that dissolve in water (Hyeon, 2013).

- Ø **Group 2: Alkaline Earth Metals:** Hydrated halides are produced when alkaline earth metals react with one another. Except for beryllium, the least metallic of the group, these halides are ionic. Halogen atoms seem to gain electrons while alkaline earth metals seem to lose them. Calcium chloride that is anhydrous absorbs enough water to dissolve its particular crystal lattice because of its strong affinity towards water. It can be created directly from limestone or by using the Solvay Process as a byproduct (Swenson, 2014).
- Ø **Group 13: The Boron Family:** Every element in Group 13 reacts with halogens to create trihalides and all boron halides are Lewis acids. Aluminum fluoride is an ionic substance with a high melting-point; aluminum halides have a dimeric structure. Al_2X_6 is the formula for the majority of aluminum halides, where X stands for chlorine, bromine, or iodine. A dimer is a molecule that develops when two AlX_3 units—or, more generally, any two units that are identic can combine. Lewis acids are extremely reactive with aluminum halides. Adducts, which are acid-base compounds that accept electrons and form, are an example:

$AlCl_3 + (C_2H_5)_2O \rightarrow Al\ (C_2H_5)_2OCl_3$

In the above reaction, $AlCl_3$ constitutes the Lewis acid while $(C_2H_5)2O$ is the Lewis base.

- Ø **Group 14: The Carbon Family:** Group 14 produces halides with the general formula MX_4 ($GeCl_4$, $SnCl_4$, $PbCl_4$, CCl_4, $SiCl_4$), but some of them can also make dihalides (MX_2). Due to the lack of vacant valence d-orbitals, carbon tetrahalides like CCl_4 and other tetrahalides can't be hydrolyzed.

1. **Silicon Halides:** Silicon and halogens react to form elements of the form SiX_4, whereby X denotes any common halogen. SiF_4 forms a colorless gas, while $SiCl_4$ is a transparent liquid, while $SiBr_4$ is a colorless liquid at room temperature. $SiBr_4$ can

only be partly hydrolyzed, whereas SiF_4 and $SiCl_4$ can be fully hydrolyzed.

2. **Metal Halides:** Group 14 includes lead and tin. Tin can be available as either SnO_2 or SnO_4. Tinstone contains the reducing agent $SnCl_2$, which is effective. In the past, SnF_2 was added to toothpaste, but NaF now takes its place (Swenson, 2014).

Ø **Group 16: The Oxygen Family:**

1. **Oxygen and Sulfur Halides:** Sulfur directly reacts with all halogens, with the exception of iodine. Sulfur hexafluoride, or SF_6, is produced when it instantly combines with fluorine to develop an inert and colorless gas. Additionally, it is capable of forming SF_4, a potent fluorinating agent. Sulfur and chlorine combine to make SCl_2, a red liquid that is used to make the toxic mustard gas. The following is the reaction: $3SF_4 + 4BCl_3 \rightarrow 4BF_3 + 3SCl_2 + 3Cl_2$

   Oxygen and fluoride combine to create the compounds OF_2 and O_2F_2. These molecules have structures that are similar to hydrogen peroxide, but they are highly reactive (Hyeon, 2013).

2. **Halogen Oxides:** Cl_2O, ClO_2, Cl_2O_4, and I_2O_5 are common halogen oxides. As demonstrated below, chlorine and oxygen are produced as a result of the vigorous reaction that chlorine monoxide, the hypochlorous acid anhydride, has with water: $Cl_2O + H_2O \rightleftharpoons 2HOCl$ (10)

When potassium chlorate and sulfuric acid react, chlorine perchlorate and chlorine dioxide are produced. Similar to the nitrogen compounds (NO_2) and (N_2O_4), these compounds are similar. When iodine pentoxide reacts with water, it forms iodic anhydride, as demonstrated: $I_2O_5 + H_2O \rightarrow 2HIO_3$ (11)

CHAPTER

9 Double Replacement Reactions

Contents

9.1. INTRODUCTION

A salt metathesis reaction, also known as a double displacement reaction, exchanges bonds between two chemical species to generate new compounds with comparable bonding affiliations. The structure shows this response: Interspecies ionic or covalent bonding. Interactions produce one product. Ancient literature used "double decomposition" often. When at least one chemical does not dissolve in the solvent, "double decomposition" is employed because ligand or ion exchange happens in the solid form of the reactant (Tang, 2022) (Figure 9.1).

Ø **Example:** AX(aq) + BY(s) → AY(aq) + BX(s).

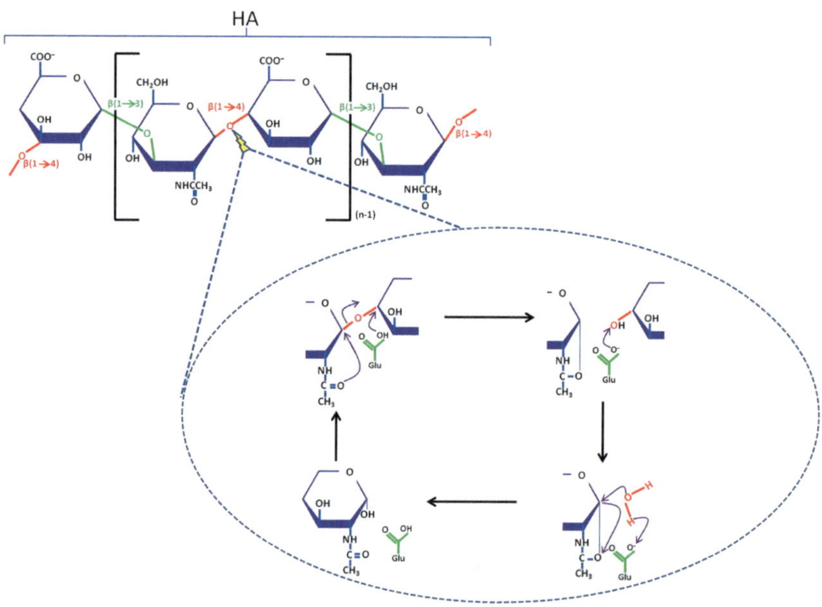

Figure 9.1. *Double displacement reaction mechanism by class I Hyals in humans.*

Source: Joy, R., Vikkath, N., & Ariyannur, P., (2018). Metabolism and mechanisms of action of hyaluronan in human biology. Drug Metabolism and Personalized Therapy. – Scientific Figure on ResearchGate. Available from: https://www.researchgate.net/figure/Double-displacement-reaction-mechanism-by-class-I-Hyals-in-humans-The-catalytic_fig4_323177082.

A type of chemical reaction called a double replacement reaction which involves the ion exchange of two different molecules. A metathesis reaction or a twofold displacement reaction are other names for this kind of reaction. These reactions typically take place in aqueous medium and are characterized by the creation of new compounds. The two compounds swap their anions (negatively charged ions) and cations (positively charged ions) during a twofold replacement process, creating two new molecules. The generic chemical equation for a reaction involving double replacement may be expressed in several ways, one of which is as follows:

AB + CD gives AD + CB.

A common form of a twofold replacement reaction is the precipitation reaction. When these reactions take place, the blending of two aqueous solutions results in the production of a precipitate, an insoluble substance. A new substance that is less soluble than the original components drive the precipitation process. The precipitate and solution can be separated effectively using filtration. A twofold replacement reaction happens when silver nitrate ($AgNO_3$) and sodium chloride (NaCl) are combined in water and contact with one another. Silver chloride (AgCl) is created when the silver cation (Ag^+), made from silver nitrate, combines with the chloride anion (Cl^-), made from sodium chloride. Due to its insoluble nature, silver chloride precipitates out of solution. The sodium cation (Na^+) in sodium chloride combines with the nitrate anion ($NO3^-$) in silver nitrate to form sodium nitrate ($NaNO_3$). The whole response can be summarized as follows: AgCl and $NaNO_3$ are created when $AgNO_3$ and NaCl mix (Tang, 2022).

The neutralization of acids and bases is a kind of twofold replacement method. An acid and a base can combine to form water and salt in the processes described below. The hydroxide anion (OH^-) from the base reacts with the hydrogen cation (H^+) from the acid to generate water (H_2O). The interaction between the anion from the acid and the base produces the salt. A twofold replacement reaction takes place when sodium hydroxide (NaOH) and hydrochloric acid (HCl) are combined. To create water (H_2O), the hydrochloric acid's hydrogen cation (H^+) must mix with sodium hydroxide's hydroxide anion (OH^-). When the sodium cation (Na^+) from sodium hydroxide and the chloride anion (Cl^-) from the hydrochloric acid mix, sodium chloride (NaCl) is the result. The full process might be demonstrated as follows: When sodium hydroxide and hydrochloric acid interact, sodium chloride and water are produced (Anscher, 2011).

Another outcome of double replacement procedures is gas generation. As a consequence of interactions like those between an acid and a carbonate or bicarbonate molecule, CO_2 gas (CO_2) is created. This kind of reaction frequently occurs when baking soda and vinegar are combined, as well as when an acid is given to an antacid pill that contains carbonate. Contrary to single replacement reactions, double replacement reactions don't need an electron transfer to take place. This is a crucial thing to think about (Adams, 2014).

9.2. A FEW TYPES OF REACTIONS

1. **Counterion Exchange:** Counterion transfer usually involves salt metathesis. Solubility charts and lattice energies can help choose reactants. HSAB theory can predict metathesis reaction outcomes. Widespread salt metathesis to make organic solvent-soluble salts. Sodium perrhenate becomes tetrabutylammonium (Anscher, 2011).

   $NaReO_4 + N(C_4H_9)_4Cl \rightarrow N(C_4H_9)_4[ReO_4] + NaCl$

   Tetrabutylammonium precipitates in water. Dichloromethane dissolves it. Salt metathesis in a nonaqueous solution converts ferrocenium tetrafluoroborate into a lipophilic salt with the tetrakis (pentafluorophenyl) borate anion.

   $[Fe(C_5H_5)2]BF4 + NaB(C_6F_5)_4 \rightarrow [Fe(C_5H_5)_2]B(C_6F_5)_4 + NaBF_4$

   $B(C_6F_5)_4$ salt persists in solution whereas $NaBF_4$ salt precipitates in dichloromethane. Metathesis reactions between inorganic salts may occur if one product is water-insoluble. Silver chloride precipitates from cobalt hexammine chloride and silver nitrate to generate the cobalt complex's nitrate salt:

   $3AgNO_3 + [Co(NH_3)_6]Cl_3 \rightarrow 3\ AgCl + [Co(NH_3)_6](NO_3)_3$

   Metathesis reactions can occur with fewer solubility reactants. Heating water, a copper(I) thiocyanate slurry, and barium hydroxide produce barium thiocyanate:

   $Ba(OH)_2 + 2CuCNS \rightarrow Ba(CNS)_2 + 2CuOH$

2. **Neutralization Reaction:** Double-replacement neutralization. An equal amount of acid and base neutralizes. This usually yields a salt-water solution. Hydrochloric acid and sodium hydroxide create sodium chloride and water.

   $HCl + NaOH \rightarrow NaCl + H_2O$

   Carbonic acid spontaneously decomposes into CO_2 and water after any acid-carbonate or bicarbonate combination. CO_2 from the reaction mélange completes it.

$CH_3COOH + NaHCO_3 \rightarrow CH_3COONa + CO_2 + H_2O$

3. **Sodium-Free Metathesis:** Silyl halides cause salt-free reductions, unlike solid salt precipitation in salt metathesis. Saltless metathesis occurs (Adams, 2014).
4. **Double Replacement Reactions:** Two new ionic compounds are created by exchanging their components. Double replacement, displacement, exchange, and metathesis reactions are other names for these phenomena. Twofold replacement response pattern. The reaction exchanges cations or anions, not both, as the initial chemical combination remains. Double replacement reactions often include ionic chemicals. Water is usually used as the solvent in double replacement (Tang, 2022).

   $BaCl_2(aq)+Na_2SO_4(aq)+BaSO_4(aq)+2NaCl(aq)BaCl_2(aq)+Na_2SO_4(aq)+BaSO_4(s)+2NaCl(aq)$

9.3. PRECIPITATION AND NEUTRALIZATION REACTIONS

Recognizing the pattern makes it easier to spot duplicate replacement replies. Recognizing common double replacement reactions can aid in prediction. Precipitation and neutralization reactions are covered on this page. In a precipitation process, two aqueous ionic compounds produce a non-water-soluble molecule. Potassium iodide interacts with lead(II) nitrate. Both are white solids that dissolve in water to produce colorless solutions. When the two obvious answers are combined, the following reaction occurs:

$Pb(NO_3)_2(aq)+2KI(aq)\rightarrow 2KNO_3(aq)+PbI_2(aqPb(NO_3)_2(\mathit{aq})+2KI(\mathit{aq})\rightarrow 2KNO_3(\mathit{aq})+PbI_2(\mathit{s})$

A golden solid was made with two easy solutions! This may be your reaction flask. A solid yellow lead (II) iodide cloud is formed by clear solutions of lead (II) nitrate and potassium iodide. The precipitate is unable to dissolve. The supernatant is the solvent and soluble component of the process. Precipitation reactions are predicted by solubility principles. The reaction is propelled forward by solid precipitates. Double replacement reactions are used in acid-base neutralization processes (Anscher, 2011).

Neutralization reaction: $HF+NaOH\rightarrow H_2O+NaF$.

$HF(\mathit{aq})+NaOH(\mathit{aq})\rightarrow H_2O+NaF(\mathit{aq})$

Water and a grain of salt are produced during the aqueous neutralization processes. The most difficult aspect of neutralization is recognizing that the reactants are acids and bases. In a neutralization reaction, if you have a strong acid or base reactant, you may typically expect a forward reaction. Double substitution equations might be molecular, full ionic, or net ionic. The molecular equation is only written in this post, although you should be familiar with the other forms.

$H_2SO+Ba(OH)_2 \rightarrow H_2SO_4(aq)+Ba(OH)_2(aq)$

Sulfuric acid, a powerful acid, is neutralized by barium hydroxide, a strong base, in two steps. What is the chemical formula for barium sulfate? According to our solubility principles, barium sulfate is insoluble and should precipitate. Our reaction is also precipitous! We may include this knowledge in our Two ionic compounds switch anions or cations during double replacement reactions. Double replacement reactions include, for example, precipitation and neutralization. Solubility rules describe precipitation reactions, which result in an insoluble product from two watery reactants (Tang, 2022).

Neutralization reactions are advantageous when the reactants are strong acids and bases. Double-replacement reactions switch the positive and negative ions of two ionic molecules to produce two new chemicals. The following equation denotes a double-replacement reaction:

$AB + CD \rightarrow AD + CB.$

Anions are BB and DD, whereas cations are AA and CC. Aqueous solutions frequently experience double-replacement reactions. A reaction needs a gas, a solid precipitate, or a molecular component like water. In a double-replacement procedure, the cations and anions from each reactant combine to create an insoluble ionic molecule. Lead (II) nitrate and potassium iodide in aqueous solutions react as follows:

$2KI(aq)+Pb(NO_3)_2(aq) \rightarrow 2KNO_3(aq)+PbI_2(s)$

Potassium nitrate can be dissolved.

The atmosphere is polluted by some double-replacement processes' gaseous wastes. Aqueous sodium chloride and hydrogen sulfide gas are produced when sodium sulfide and hydrochloric acid are mixed (Anscher, 2011).

$Na_2S(aq)+2HCl(aq) \rightarrow 2NaCl(aq)+H_2S(g)$

A molecular molecule is created as a result of another double-replacement process. Water-producing reactions are rather common. Aqueous hydrochloric acid and sodium hydroxide combine to produce water and sodium chloride:

$HCl(aq)+NaOH(aq) \rightarrow NaCl(aq)+H_2O(l)$

Make a balanced chemical equation for these double-replacement reactions.

- $NaCN(aq)+HBr(aq) \rightarrow NaCN(aq)+HBr(aq) \rightarrow$ (hydrogen cyanide gas is formed).
- $(NH_4)_2SO_4(aq)+Ba(NO_3)_2(aq) \rightarrow (NH_4)_2SO_4(aq)+Ba(NO_3)_2(aq) \rightarrow$ (a precipitate of barium sulfate forms).

1. **Solution:** Plan the problem first. Production capacities for gas the response of A. B is triggered. Use the ionic charges of the reactants in both instances to produce the products (Tang, 2022).
2. **Solve:**

i. There are 11 anions and +1+1 cations in both reactants. Balanced after changing partners:

$NaCN(aq)+HBr(aq) \rightarrow NaBr(aq)+HCN(g)$

ii. Barium and sulfate have the formulae 2+2+ and 22, whereas ammonium and nitrate have the formula 1+1+. Exchange partners and come up with fresh formulations with this in mind. The equation balances out (Adams, 2014).

$(NH4)_2SO_4(aq)+Ba(NO_3)_2(aq) \rightarrow 2NH_4NO_3(aq)+BaSO_4(s)(\ldots$

3. **Step 3: Assess Your Findings:** There are two alternate responses. Formulae that are accurate and equations that are balanced. Gas and molecule-producing reactions are possible. Aqueous sodium chloride, CO_2, and water are the products of the reaction between hydrogen chloride and sodium carbonate (Anscher, 2011).

$Na_2CO_3(aq)+2HCl(aq) \rightarrow 2NaCl(aq)+CO_2(g)+H_2O(l)$

Sodium chloride and silver nitrate produce a two fold replacement reaction in water. Silver nitrate and sodium chloride are examples of ions. In water, both reactants produce ions. The sodium ion picks up the nitrate anion to create sodium nitrate, whereas the silver ion takes up sodium chloride to create silver chloride. Additionally, ionic compounds are created. Silver chloride precipitates due to the low solubility of silver in water (Tang, 2022).

$AgNO_3 + NaCl \rightarrow AgCl + NaNO_3$

Barium sulfate and sodium chloride are produced when barium chloride and sodium sulfate are combined.

$BaCl_2(aq) + Na_2SO_4(aq) \rightarrow BaSO_4(s) + 2\ NaCl(aq)$

All reactions include the utilization of ionic reactants and products. Barium sulfate solution solidifies substitute on both sides Reactions is identified Cations replacing anions in a chemical equation indicates a twofold replacement reaction. A reaction between two aqueous solutions is desired if it produces both an aqueous product (aq) and a solid product (s) that precipitates. Precipitate is expected to result from a twofold replacement reaction when unidentified reactants are mixed (Anscher, 2011) (Figure 9.2).

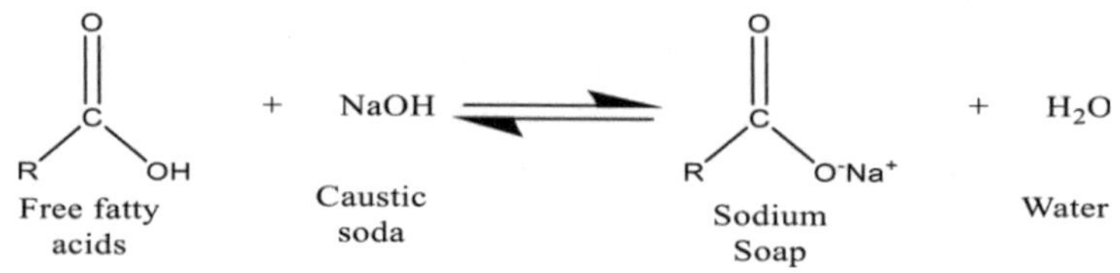

Figure 9.2. *Neutralization reaction of free fatty acids with caustic soda.*

Source: Ombick, G., Mengata, M., Ghislain, Tchamdjio, N., Emeric, & Adolphe, M. I., (2022). Investigation of the thermal aging of the natural monoester/paper mixed insulation. Electrical Engineering. – Scientific Figure on ResearchGate. Available from: https://www.researchgate.net/figure/Neutralization-reaction-of-free-fatty-acids-with-caustic-soda_fig1_360635020.

Solubility criteria can forecast whether the reactants will dissolve and create a precipitate (indicating a double replacement reaction) if the reaction cannot be observed visually. Examples of double replacement processes include neutralization, alkylation, acid-carbonate reactions, counter-ion exchange, precipitation reactions, and double decomposition reactions (Adams, 2014).

Neutralization and precipitation reactions rule general chemistry:

- Both bases and acids can be neutralized in two steps. Normally, silver dissolves in water to form salt. Strong acids and bases generate forward reactions.
- Sodium hydroxide and hydrofluoric acid combine in water to form water and sodium fluoride. While sodium hydroxide is naturally basic, hydrofluoric acid is acidic by nature. Overall impression:

Acid + base → water + salt

In this case, the reaction is:

$HF(aq) + NaOH(aq) \rightarrow H_2O + NaF(aq)$

The baking soda volcano balances out the vinegar and baking soda.

Carbonic acid (H_2CO_3) and sodium acetate (CH_3COONa) are produced during the neutralizing process before CO_2 and sodium carbonate.

$$NaHCO_3 + CH_3COOH(aq) \rightarrow H_2CO_3 + CH_3COONa$$

Complex formulations hide the fact that cations exchange anions. The atoms of the anions of the reactants and products exhibit twofold replacement. Two aqueous ionic compounds precipitate an insoluble ionic product. Lead(II) nitrate and potassium iodide react to produce potassium nitrate and (insoluble) lead iodide (Anscher, 2011).

$$Pb(NO_3)_2(aq) + 2\ KI(aq) \rightarrow 2\ KNO_3(aq) + PbI_2(s)$$

The chemical formula's (s) stands for precipitation. Lead iodide is the precipitate, and water, soluble reactants, and product are the supernatant. As the product leaves the solution, precipitation speeds up the process. Double displacement reactions are bimolecular processes that change the compound reactants AB and CD into AC and BD. These reactions are sometimes referred to as metathesis reactions. This results from one group attacking another nucleophilically, followed by the release of another group to create a new molecule. This process repositions fragments between alkenes around double bonds during olefin metathesis (Tang, 2022).

9.4. ENZYMATIC DOUBLE-DISPLACEMENT REACTIONS

When products are released before all substrates bind to the enzyme, a double displacement reaction occurs. Double displacement reactions are referred to as "ping-pong" reactions. The enzyme bounces substrates on a table like a ping-pong ball, thus the name. Aspartate aminotransferase undergoes a two-step displacement reaction. This enzyme transforms aspartate to alpha-ketoglutarate, which results in the formation of oxaloacetate and glutamate. This response is not sequential. Aspartate binds to the enzyme and donates its amino group, resulting in the formation of a substituted enzyme intermediate. After donating the amino group, aspartate produces oxaloacetate. The amino group is acquired by alpha-ketoglutarate, the enzyme's second substrate, from the substituted enzyme intermediate. Glutamate is released throughout the process. To occur, a double chemical reaction must produce one of three products: solid, water, or gas. Water is created here. All double replacement procedures need two aqueous (water-soluble) reactants. You have a bucket of ions before the reaction. The objective is to rearrange the ions into new compounds while avoiding the formation of similar-charged ions. Water

(H_2O) is a molecular compound, but we conceive of it in terms of two ions: hydronium (H^{+1}) and hydroxide (OH^{-1}). The compound would be the HOH. Thus, one reactant must include hydroxide (OH^{-1}) and the other must have hydronium (H^{+1}) ion. In a bucket, these reactants make water (H_2O) (Tang, 2022).

Consider the following example. Because these reactions are so common, we refer to the reactants and chemical reaction as:

- Acd. degrades in water into ions, notably hydronium (H^{+1}). Acids are compounds H-ionic. HNO_3 separates into H^{+1} and NO_3^{-1} in Example #1. #2: H_2SO_4 decomposes into two H^{+1} and one SO_4^{-2} (Adams, 2014).

In water, bases degrade into ions, notably the hydroxide ion (OH^{-1}).

- Neutralization R—a two-replacement reaction that produces H_2O. Reactants that are acidic and basic.

Ø **Examples of Neutralization Reaction:**

$H_2O(l)$ (or 1 $HOH(l)$) + $NaCl(aq)$ ==> 1 $NaOH(aq)$ + 1 $HCl(aq)$ $2H_2O$ (or 2 HOH) + $CaCl_2$ (aq) ==> 1 $Ca(OH)_2$ (aq) + 2 HCl

Neutralization ionic net reaction. To locate any net ionic reaction, type (net ionic reactions).

Determining neutralization reactions and net ionic reactions:

Ø **Step 1:** 2 $H_2O(l)$ + $CaCl_2$ (aq) ==> 1 $Ca(OH)_2$ (aq) + 2 $HCl(aq)$ (Balanced Molecular Equation)

Ø **Step 2:** $1Ca^{+2}$, $2OH^{+1}$, $2H^{+1}$, $2Cl^{-1}$ → $1Ca^{+2}$ + $2Cl^{-1}$ + $2H_2O(l)$

Ø **Step 3:** I am unable to cancel spectator ions on this website. Excellent work, top. Ca^{+2} and 2 Cl^{-1} spectator ions are present on both sides of the arrow.

Ø **Step 4:** 2 OH^{-1} + 2 H^{+1} = 2 $H_2O(l)$. However, you may obtain the Net Ionic Reaction by canceling a "2" in each coefficient.

$H_2O(l) = OH^{-1} + H^{+1}$.

In a twofold replacement process, two ionic compounds swap ions to produce two new compounds containing the same ions. Important takeaways in a double replacement reaction, two reactants swap cations or anions to form two new products (Adams, 2014).

Double displacement reactions, also known as metathesis reactions, are double replacement processes. The processes of neutralization, precipitation, and gas generation are all double replacement processes.

A+B- + C+D- A+B- + C+D

By swapping the positive-charged cations and negative-charged anions of the reactants, this reaction produces two new products. Also known Metathesis and double replacement reactions are other names for double displacement processes (Anscher, 2011).

The answer:

It's a two-pronged replacement reaction. Silver's nitrite was substituted for sodium's chloride. Sodium chloride and hydrogen sulfide are formed when sodium sulfide and hydrochloric acid react:

$NaCl+H_2S = Na_2S+HCl$

Neutralization, precipitation, and gas generation are examples of metathesis reactions. The neutralization process is an acid-base process that results in a neutral pH solution. A precipitate is formed when two substances react. The precipitate is either insoluble in water or sparingly soluble in water. Gas is produced via gas production processes. Previously, the gas manufacturing method produced hydrogen sulfide. In a double-replacement or double-displacement reaction, two reactants swap ions to produce two new molecules. The double-replacement process commonly precipitates in an aqueous solution. Like ionic molecules, acids, and bases undergo double-replacement processes. Product compounds, like reactant molecules, form bonds (Tang, 2022).

9.4.1. General Formula

The response of double replacement is AB + CD gives AD + BC. Real-world reactions must be balanced. Aqueous solutions are frequently abbreviated as aq., and precipitates as ppt. Their chemical behavior is characterized by double-replacement reactions for ionic compounds. Ionic substances exchange cationic and anionic ions. Two brand-new ionic compounds have been developed (Anscher, 2011).

9.4.2. How Does a Double Replacement Occur and What Causes It?

In a double-replacement reaction, the cations or anions of two ionic compounds are exchanged. The interaction of two substances in an aqueous solution can yield water and an insoluble precipitate, or a gas. The chemical reaction that occurs when two compounds exchange ions to form two new compounds is known as a double replacement reaction. Cation and anion transfer is a straightforward and uncomplicated method for determining whether a twofold displacement reaction has occurred. When the states of matter are specified, it is best to search for watery reactants and a singular solid product because the reaction frequently precipitates. Double replacement reactions are specific type of chemical reactions in which two ionic molecules combine to form a new molecule. In the presence of iron(III) bromide ($FeBr_3$) and sodium hydroxide (NaOH), a bimolecular substitution reaction takes place. Consequently, NaBr and $Fe(OH)_3$ precipitate. The reaction is depicted by the chemical equation below: Aqueous ferric bromide ($FeBr_3$), sodium bromide (NaBr), and solid ferric hydroxide ($Fe(OH)_3$) precipitate and react with one another in a 1:1:3:3 molar ratio (Tang, 2022).

How does one achieve equilibrium in a double-replacement reaction? Barium nitrate and ammonium sulfate endure a chemical reaction. Ammonium nitrate (NH_4NO_3) and barium sulfate ($BaSO_4$) are the process's ultimate chemical compounds. In the reaction between aqueous $Ba(NO_3)_2$ and $(NH_4)_2SO_4$, $BaSO_4$, and aqueous NH_4NO_3 are produced. The perspective expressed in the subsequent remark is excessive. There appears to be a transition between the sulfate ($SO4^{2-}$) and nitrate (NO^{3-}) anions in the compounds. Two ions are present on the left side of the chemical equation, but only one is present on the right. The equation becomes balanced when NH_4NO_3 is multiplied by a factor of two. The reaction between $Ba(NO_3)_2$ and NH_4NO_3 precipitates solid $BaSO_4$ and generates liquid NH_4NO_3 (Anscher, 2011).

9.5. PRINCIPLES OF DOUBLE REPLACEMENT REACTION

The law of solubility applies to all chemical reactions involving double replacement. This method allows for the forecasting of precipitation. Deposition of solid precipitation powers the process. Precipitation forecasting utilizes the solubility table. Reactions involving the exchange of two chemical species. According to published research work, there are two primary categories of double-replacement responses. The phenomenon of precipitation occurs when two ionic compounds dissolve in an aqueous medium and then react to produce a third component. As a consequence of the solvent, the new chemical solidifies. A precipitate facilitates the reaction's catalysis (Adams, 2014).

- Ø ***Examples:*** A precipitation reaction between $AgNO_3$ and NaCl in the presence of sodium nitrate solution produces silver chloride.

In an aqueous solution, silver nitrate ($AgNO_3$) and sodium chloride (NaCl) react to precipitate solid silver chloride (AgCl). The reaction between sodium chloride (NaCl) and sodium sulfate (Na_2SO_4) produces sodium chloride (NaCl) and a white precipitate of barium sulfate ($BaSO_4$)(aq.) We are contemplating the chemical formula Na_2SO_4 + (aq) $BaCl_2$. PbI_2, an insoluble compound in KNO_3, is produced by the chemical reaction between potassium iodide (KI) and lead (II) nitrate ($Pb(NO_3)_2$). According to the chemical equation, the product of the reaction between $Pb(NO_3)_2$ and 2KI(aq) is $Pb(NO_3)_2$ (aq).

Interactions between acids and bases result in neutralization reactions. The reactions between aqueous solutions of acids and bases frequently produce salt and water. The chemical reaction between sulfuric acid (H_2SO_4.) and sodium hydroxide (NaOH) produces sodium sulfate (Na_2SO_4) and water (H_2O). As a result of the chemical reaction between sodium hydroxide (NaOH) and hydrochloric acid (HCl), sodium chloride (NaCl) and water (H_2O) are generated. The chemical equation NaCl + H_2O (l) = HCl + NaOH (aq.) describes the reaction between sodium chloride and water in liquid form to produce hydrochloric acid and sodium hydroxide in aqueous form. When vinegar and baking soda are combined, carbonic acid (H_2CO_3) and sodium acetate (CH_3COONa) are generated. The chemical equation describes the reaction involving H_2CO_3, CH_3COONa, $NaHCO_3$ (in aqueous form), and CH_3COOH. Carbonic acid is believed to be an unstable substance. CO_2 is emitted during the decomposition process, causing the formation of bubbles. Water (H_2O) and CO_2 react to form aqueous carbonic acid (H_2CO_3), as depicted by the user's chemical equation. The double replacement reaction is operating again (Tang, 2022).

The phenomenon of gas formation occurs when gas is produced as a product of a chemical reaction. The gaseous components of the solution rise into the atmosphere of the planet. When Na_2S-containing HCl solutions are mixed with Na_2S-containing HCl solutions, NaCl and H_2S gas are produced. ZnS (zinc sulfide) dissolves in hydrochloric acid (HCl) to produce $ZnCl_2$ and H_2S gas. The chemical equation represents the reaction between two molecules of sodium chloride (NaCl) in an aqueous solution and gaseous hydrogen sulfide (H_2S). The reaction between solid zinc sulfide (ZnS) and hydrochloric acid (HCl) produces gaseous hydrogen sulfide (H_2S) and aqueous zinc chloride ($ZnCl_2$), as represented by the following chemical equation:

$$H_2S(g) + ZnCl_2 = ZnS(s) + 2HCl + H_2S(g)$$

The double-replacement response phenomenon occurs frequently in both natural and anthropogenic environments. Baking soda and vinegar interact in advantageous ways for human(Tang, 2022).

9.6. PREDICTION OF DOUBLE REPLACEMENT REACTIONS

Predicting double displacement responses is challenging. Precipitation is a two-step substitution process that occurs in the environment. For accurate precipitation episode prediction, it is necessary to comprehend the charge balances of the two newly-formed species. Similar to single displacement reactions, double displacement reactions do not always exhibit responsiveness. Therefore, the following section will discuss reactions that occur to precisely ascertain product charge equilibrium. To predict a double replacement reaction accurately, it is advised to follow the next three steps. The anticipated reaction will be followed by the addition of a subsequent phase. By altering the reactant companions of cations and anions, it is possible to produce new compounds. Adjust the chemical composition of the product to account for the presence of ionic charges. The equation must be equitable. Determine what will occur when two Li_2SO_4 and $BaCl_2$ aqueous solutions are combined. The reactant side must change the partners of anions and cations to form new molecules (products). In this stage, just the oxygen moves from reactant to product. Sulfate oxygen is polyatomic. Lithium and chloride are not shown here. The ion has no subscripts. Instead, they show how much of that ion-maintained charge equilibrium. Ion charges determine product compositions. Ions are in balance; thus no changes are needed. The charges of $Ba^{2+}Ba^{2+}$, $Cl^{-}Cl^{-}$, $Li^{+}Li^{+}$, and SO_4^{-2} will verify this. (Note: Please ask us how we computed these costs. The assumption that all coefficients are one was not modified since the anion and cation charges were canceled in each compound. After the product formulae are precisely anticipated, we should balance the equation. We may estimate product coefficients by examining the beginning subscripts. Subscripts should have been identified before. The coefficients below are added to unbalanced equations (Anscher, 2011).

$BaCl_2(aq)+Li_2SO_4(aq) \rightarrow BaSO_4+2LiCl$

In a twofold displacement process, the anions and cations of two compounds swap locations to generate two different molecules. These reactions are usually: The SO_4^{2-} anion replaces the $2Cl^{-}$ anion in barium chloride ($BaCl_2$) and magnesium sulfate ($MgSO_4$), becoming $BaSO_4$ and

$MgCl_2$. Another twofold displacement process involves potassium iodide and lead(II) nitrate: Le Chatelier's Principle states that reactions can go either way until equilibrium [22]. Spontaneous responses reach equilibrium. If they occur at constant temperature and pressure, their negative Gibbs free energy will lower the processes. They move easily. Direct reactions dominate. Examples:

Hydrogen and oxygen produce water.

$2H_2 + O_2 \rightleftharpoons 2H_2O$

- Water with acetic acid generate acetate and hydronium ions. Positive equilibrium-seeking responses are non-spontaneous. They raise the reaction's Gibbs free energy under constant temperature and pressure. Progress requires work.
- Charging an electrolytic cell-based DC battery using an external power source.
- Photosynthesis uses electromagnetic energy, usually sunshine.

Photons, CO_2, and H_2O are elementary particles. [CH_2O] oxygen sugar. An element or molecule combines with an oxidantusually oxygen—to produce heat or light in a combustion processes. Combustion uses hydrocarbons. About 1 mole (114 g) of octane burning in oxygen release 5500 kJ energy. Carbon, magnesium, and sulfur may burn with oxygen. When referring to a chemical, something being reactive suggests that it participates in chemical reactions easily and quickly. Some metals are significantly more reactive than others when compared to others. It is possible to identify the order in which metals are more reactive to other elements. A reactivity series is a sort of sequence like this. Nonreactive metals either engage in chemical processes less actively or do not react at all. Metal and the compound of another metal are both necessary for displacement reactions to occur. A metal with a higher degree of reactivity will displace or drive a metal with a lower level of reactivity out of its compound in a displacement reaction. The metal with the lesser degree of reactivity stays uncombined after the reaction. Chemical interactions no longer link it to any other elements. It has reached its essential condition (Tang, 2022).

Ø ***Example:*** Copper's reactivity is lower than that of magnesium. When a piece of magnesium is dipped into a solution of blue copper sulfate, a displacement reaction occurs.

After the magnesium pushes the copper out of the way, the reaction produces copper and a solution of magnesium sulfate.

The mathematical formula is as follows: When magnesium sulfate is combined with copper, copper sulfate is formed.

The equation can be represented by the symbol:

$MgSO_4$ (s) + Cu (s) → Mg(s) + $CuSO_4$ (aq).

The inverse of this reaction can never occur. Copper cannot displace magnesium from a complex because it is a less reactive than magnesium.

Metals may be extracted from their oxides via displacement techniques. The utilization of displacement processes can be used to extract metals from their oxides.

Carbon is a non-metallic element that is abundant and affordable. In compared to other metals such as copper, iron, and lead, it has a higher degree of reactivity. This suggests that carbon might be utilized to extract some metals from the metal oxides in which they are located (Anscher, 2011).

The powdered rock is mixed with carbon and then heated to extreme temperatures. As an example, copper oxide + carbon → copper + carbon dioxide

Some metals, such as aluminum, are incompatible with this method due to their greater reactivity than carbon. Carbon is unable to replace these metals (Tang, 2022).

Displacement reactions occur in nonmetallic materials. Iodine, bromine, and chlorine all quite similar exhibit reactions. Iodine has the lowest reactivity among the elements, whereas chlorine has the highest. Chlorine can displace bromine and iodine from their complexes in these nonmetals.

- Bromine, unlike chlorine, can displace iodine.
- Iodine cannot be used to replace chlorine or bromine. As iodine is least reactive.
- When chlorine is added to a sodium bromide solution, a reaction occurs, and the following stages occur:

Bromine + Sodium Chloride Chlorine + sodium bromide

- When bromine comes into contact with a sodium iodide solution, the following reaction occurs bromine and sodium iodide mix to generate iodine and sodium bromide.

When iodine is added to either sodium chloride or sodium bromide, no reaction occurs. Iodine lacks the reactivity required to remove those components from the system (Anscher, 2011).

A single replacement reaction is a sort of chemical reaction in which a single uncombined ingredient replaces another element already present in a compound; in other words, one element in the compound swaps places with another element in the compound. These reactions can be classified into the following categories:

When magnesium is injected into the water instead of hydrogen, a single displacement reaction occurs, resulting in the formation of magnesium hydroxide and hydrogen gas. The anions and cations of two independent compounds exchange places during a twofold replacement process, resulting in the production of two wholly unique molecules. These answers are best described by the following categories: For example, the interaction of barium chloride ($BaCl_2$) and magnesium sulfate ($MgSO_4$) results in the exchange of the anion with the charge of for the anion with the charge of $SO4^{2-}$, yielding the molecules $BaSO_4$ and $MgCl_2$. Another example of a twofold displacement reaction is the interaction of potassium iodide with lead (II) nitrate to create both lead (II) iodide and potassium nitrate. The main characteristics of reactions involving double replacement are as follows:

- Ø **Ionic Structure:** Double replacement processes occur in aqueous solutions containing two distinct ionic molecules. As new molecules are formed, cations, also referred to as positive ions, and anions also referred to as negative ions, switch places (Tang, 2022).

9.7. DIFFERENCES BETWEEN SINGLE REPLACEMENT REACTIONS AND DOUBLE REPLACEMENT REACTIONS

1. **Reactivity:** Reactivity develops when a less reactive element is supplanted with a more reactive one. Ion exchanges are necessary for double replacement procedures involving two molecules. These categories both contain reactive reactions.
2. **Ionic Nature:** Double replacement reactions between two distinct ionic molecules are less common than single replacement reactions between a metal and an ionic molecule. The complete composition of an ion.
3. **Redox Reactions:** Electron exchange-containing redox reactions are a subset of redox reactions. Sometimes, these reactions are also referred to as single-substitution reactions. In double

replacement techniques, electron transport plays no function.

4. **Products:** In single replacement reactions, a novel compound is always produced alongside a free element or ion. Two entirely novel molecules are generated as a result of the two replacement processes (Anscher, 2011).
5. **Precipitation and Gas Formation:** Emissions of pollutants and precipitation Unlike single replacement reactions, double replacement reactions can produce gases and precipitates. Because double replacement reactions can produce these metabolites, single replacement reactions manifest these characteristics infrequently.
6. **Complete Ion Exchange:** Every single ion transfer between reactants. In contrast to double replacement reactions, single replacement reactions do not require the complete exchange of all ions between reactants. Single replacement reactions, on the other hand, involve the substitution of a single molecule element (Tang, 2022).
7. **Number of Reactants:** There are always two reactants in single replacement reactions, one of which is typically a complex and the other a purified element. They are represented by the letters A and BC, respectively. In double replacement procedures, however, two distinct chemicals (AB and CD) are required as reactants.
8. **Reaction Conditions:** In the presence of a liquid phase or solvent, such as water or an organic solvent, single replacement reactions are common. Consider the following solvents as an example. Sometimes, these interactions can take on visible manifestations. Double replacement reactions are prevalent in aqueous solutions because the reactants for this type of reaction must be dissolved in water for it to occur.
9. **Catalysts:** Single-replacement processes can benefit from catalysts, which accelerate a reaction without being squandered. The process that converts hydrogen gas (H_2) and chlorine gas (Cl_2) into hydrogen chloride (HCl) is catalyzed, for instance, by sunlight. This method produces hydrogen chloride. Typically, catalysts are not necessary for double replacement procedures. This is because spontaneous ion exchange can occur whenever sufficient reactants are present (Anscher, 2011).

10. **Stoichiometry:** The stoichiometry of reactions involving a single replacement can be calculated by evaluating the balanced equation. The quantity of product produced is directly proportional to the purified element to complex molar ratio. Both the balancing equation and the molar ratios of the ions involved can be used to characterize the stoichiometry of double replacement processes.
11. **Reactivity Series:** The activity series, also called the reactivity series, is a classification and ranking system for metals based on their reactivity. This technique influences the utilization of single-replacement operations. The reactivity series is one of the method for determining whether one metal can substitute for another in a complex. Because double replacement reactions involve the transfer of ions between molecules, the reactivity series has no direct effect on them (Tang, 2022).
12. **Observability:** Typically, single replacement reactions result in observable modifications such as color changes, gas emissions, or solid precipitation. Due to the possibility of precipitation resulting from these activities, double replacement reactions, also known as precipitation reactions, may occur. If certain assays or markers are not utilized, it is possible that double replacement reactions will not reveal discernible alterations.

9.8. APPLICATIONS OF DOUBLE REPLACEMENT REACTIONS

1. **Precipitation Reactions:** Precipitation-inducing reactions are a type of double replacement reaction that generates an insoluble product (precipitate). Combining two aqueous solutions produces a solid substance, demonstrating the existence of a twofold substitution process. One method for isolating the precipitate from the solution is filtration.
2. **Acid-Base Neutralization Reactions:** In a double replacement reaction, an acid and a base react to produce a reaction. The process produces byproducts of water and salt salinity. Acid-base neutralization mechanisms are readily identifiable due to their official name. The interaction of hydrogen ions (H^+) from the acid and hydroxide ions (OH^-) from the base produces water. The interaction of positive ions from the acid and negative ions from the base produces salt. Some double replacement reactions

can lead to the creation of water and a salt. Both an acid and a base must be present for these reactions to occur. Many different applications, including the creation of salts, the creation of medications, and the neutralization of acidic or basic solutions, frequently make use of these reactions (Tang, 2022).

3. **Gas Formation:** Double replacement strategies that have the potential to produce gas may result in gas production. When an acid reacts with a carbonate or bicarbonate molecule, CO_2 gas is generated. This type of reaction is common when baking soda and vinegar are combined, or when an acid is added to a carbonate-containing antacid tablet. When an antacid tablet is pulverized and administered to a child, a similar effect occurs. As a consequence of some kinds of double replacement processes, gases are created. For instance, when an acid reacts with a carbonate or bicarbonate molecule, CO_2 gas is created as a byproduct. Numerous products that need the production of CO_2, such as baking powder and effervescence tablets, use this gas evolution technique (Anscher, 2011).
4. **No Redox Reactions:** Double replacement processes do not generate redox reactions because, unlike single replacement processes, they do not require any transfer of electrons. The oxidation states and oxidation indices of the elements were discovered to be unchanged.
5. **Complete Ion Exchange:** Instead of them, ions are utilized. During a procedure known as a twofold replacement, the ions in the two reactants endure a complete transformation. Positive ions in one substance exchange positions with positive ions in another substance to create two new molecules. This is the outcome of the procedure (Tang, 2022).
6. Double replacement reactions commonly result in the formation of a solid precipitate when two aqueous solutions are mixed. Precipitation reactions are what these reactions are called. Several analytical techniques, including qualitative analysis, take advantage of this feature so that specific ions in a solution may be identified by the distinctive precipitates they leave behind.
7. Double replacement reactions, which are crucial to the process, are frequently utilized in the water purification process. For instance, when calcium hydroxide (lime) is added to water that

already contains dissolved metals like lead or copper, double replacement processes occur in the water. Insoluble metal hydroxide precipitates are produced as a result of these reactions, and they may be removed from the water by filtration.

8. Making insoluble Compounds Both in the laboratory and in the workplace, double replacement processes are utilized to create insoluble chemicals. Both of these responses could be present. By combining and mixing two aqueous solutions that contain the required ions, scientists can precipitate certain compounds that are otherwise challenging to get. The production of pigments, catalysts, and a variety of other materials typically uses this method (Anscher, 2011).

9. Metathesis is a step in the synthesis of chemical compounds. Double replacement reactions, which are useful in organic chemistry, can be used to synthesize a variety of organic compounds. Organic molecules engaged in metathesis processes may create new carbon-carbon or carbon-heteroatom bonds. This would enable the development of more intricate biological structures.

9.9. DISADVANTAGES OF DOUBLE REPLACEMENT REACTIONS

- The majority of double replacement reactions occur in aqueous solutions when the reactants are dissolved in water. They might not exist in other solvent systems or have limited application. In certain experimental conditions, this constrains the magnitude of these reactions.
- Double replacement reactions may be relatively slow in comparison to other forms of reactions. The exchange of ions between reactants requires time, and concentration, temperature, and the presence of catalysts can all affect the rate of the reaction. In exceptional circumstances, the reaction may require higher temperatures or extended reaction times to complete.
- Rarely, double replacement reactions can produce undesirable metabolites or side products. These side reactions may reduce the overall efficacy of the reaction and confound the purification process. To prevent the formation of undesirable compounds, it

is essential to meticulously control the reaction conditions and select suitable reactants (Tang, 2022).

- Their dependence on product solubility. If the products are highly soluble in the reaction medium, they may remain in the solution instead of precipitating out as expected. This may make it difficult to isolate and purify the required product which is needed.
- The completion of double replacement reactions is not always guaranteed. Factors such as the solubility of the reactants and products, the formation of complexes, and the presence of competing equilibria can influence the extent of the reaction. Incomplete reactions may lead to decrease yields and impurities in the final product.
- Additional ions or compounds can make double replacement processes sensitive to reaction parameters including pH, temperature, and the presence of additional ions or compounds. Variations in these variables may affect the reaction's outcome, resulting in contradictory findings. To obtain the desired yields and selectivity, it is necessary to meticulously optimize and control the reaction parameters.
- Since double replacement reactions frequently involve the exchange of multiple ions, multiple products may be formed. Particularly when there are multiple competing reactions or when the reactants have comparable reactivity, it can be difficult to obtain high selectivity for a single product. The production of selective products may necessitate the use of specific catalysts or tailored reaction conditions (Anscher, 2011).

In conclusion, for the complicated formation steps to be completed, a metal ion and a ligand are needed. A complex ion is produced as a result of this. When a ligand attaches to a metal ion, a coordination complex is created. A ligand may be an ion or a molecule. These reactions often occur in aqueous solutions and are commonly associated with transition metal ions. The reaction that occurs when silver nitrate ($AgNO_3$) is exposed to ammonia (NH_3) is an example of this: $AgNO_3$ + $2NH_3$ gives $[Ag(NH_3)_2]^+$ + NO^{3-}. This reaction results in the creation of a silver-ammonia complex. The silver ion accepts the ammonia ligand (Tang, 2022).

In double replacement reactions, new compounds are created by exchanging the positive ions (cations) and negative ions (anions) from the reactants. This is an important information to remember. The reactions are driven by the formation of insoluble products (precipitates), neutralization of acids and bases, gas evolution, or synthesis of stable complex ions. Double replacement reactions have a wide range of applications in several fields. They are used for qualitative analysis in analytical chemistry to identify specific ions based on the distinctive precipitates they form. By examining the precipitate that is produced and making observations, chemists can determine which ions are present in a solution. One technique for removing contaminants from water during the treatment process is double replacement reactions. For instance, calcium hydroxide, often known as lime, can produce insoluble metal hydroxide precipitates that can be removed by filtering when added to water that already contains dissolved metals like lead or copper (Anscher, 2011).

Double replacement processes are engaged not only in the synthesis of pigments, catalysts, and other materials. If they carefully select the reactants, scientists can create compounds that are insoluble by the use of double replacement processes. These substances have a wide range of possible applications in many different industries, such as the production of paint, the synthesis of catalysts, and the study of materials. In a nutshell, double replacement reactions are defined by the ion exchange between two distinct chemicals, which eventually leads to the creation of two new molecules. Different types of double replacement reactions include precipitation processes, acid-base neutralization events, gas-forming reactions, and complex creation processes. These procedures have major applications in the purification of water, the manufacture of a wide range of materials, and other branches of analytical chemistry as well. (Tang, 2022).

CHAPTER

10 Redox Reactions

Contents

10.1. INTRODUCTION

Redox reactions, short for reduction-oxidation reactions, are chemical processes involving electron transfer between species. In these reactions, one substance undergoes oxidation by losing electrons, while another undergoes reduction by gaining those electrons. The term "redox" combines the words "reduction" and "oxidation" to represent the dual nature of these reactions in short.

Redox reactions are of utmost importance in the field of chemistry. They are involved in various essential chemical processes, such as energy production, electron transfer, and the synthesis of many compounds. Redox reactions are crucial in understanding and analyzing chemical reactions and their mechanisms. Additionally, they provide a foundation for studying more complex chemical processes, including electrochemical cells and biological redox reactions (Felton, 2012) (Figure 10.1).

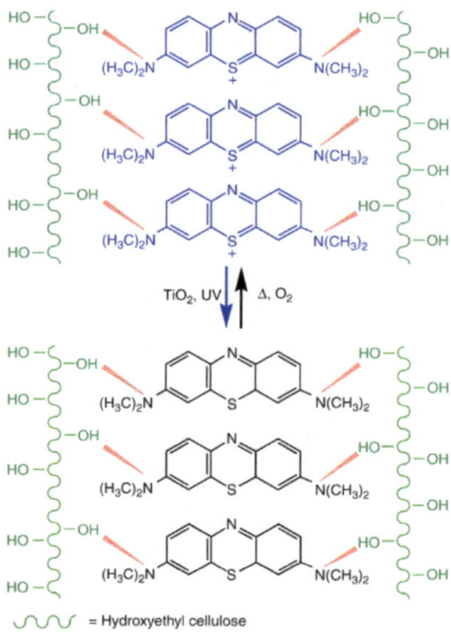

Figure 10.1. *Reversible redox reactions involved in the color switching of a TiO_2/MB/HEC composite film.*

Source: Wang, W., Xie, N., He, L., & Yin, Y., (2014). Photocatalytic color switching of redox dyes for ink-free light-printable rewritable paper. Nature Communications. – Scientific Figure on ResearchGate. Available from: https://www.researchgate.net/figure/Reversible-redox-reactions-involved-in-the-colour-switching-of-a-TiO2-MB-HEC-composite_fig2_269175728.

Oxidation and reduction are fundamental concepts in redox reactions. Oxidation refers to the loss of electrons by a substance, leading to an increase in its oxidation state. It is often accompanied by a gain of oxygen or the loss of hydrogen atoms. On the other hand, reduction involves a substance's gain of electrons, decreasing its oxidation state. It is frequently associated with the gain of hydrogen or the loss of oxygen atoms (Miura, 2015).

Understanding the principles of oxidation and reduction is crucial for comprehending redox reactions. The transfer of electrons from oxidized to reduced species drives these reactions. By studying the mechanisms and properties of oxidation and reduction, chemists can gain insights into the behavior of various substances and predict their reactivity in redox processes (Felton, 2012).

10.2. OXIDATION AND REDUCTION

Oxidation is a chemical process characterized by losing electrons from a substance. It involves the increase in the oxidation state of an atom, ion, or molecule. Oxidation often occurs simultaneously with the addition of oxygen or the removal of hydrogen atoms from a compound. However, it is important to note that oxidation can also occur without oxygen, as the concept of oxidation extends beyond oxygen's role in these reactions. In essence, oxidation represents the transformation of a species into a more positive or less negative state by losing electrons (Miura, 2015).

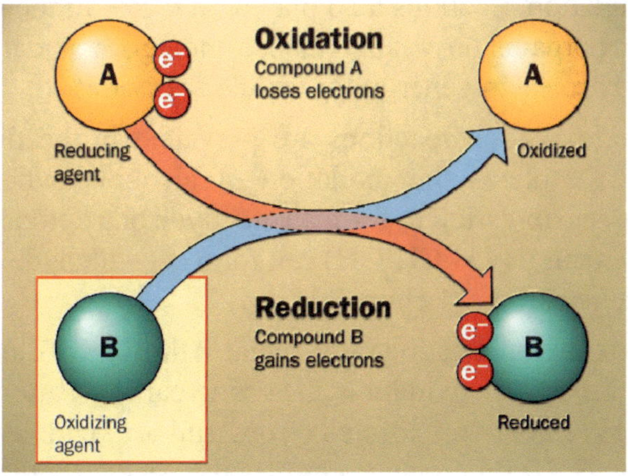

Figure 10.2. *Basic oxidation and reduction.*

Source: Azman, N., Ramli, M., & Isa, S., (2019). A review of hybridization of carbon nanotube into graphene for gas sensor application. IOP Conference Series: Materials Science and Engineering. – Scientific Figure on ResearchGate. Available from: https://www.researchgate.net/figure/Basic-oxidation-and-reduction-7_fig1_335179045.

Oxidation reactions can occur in various chemical contexts and are not limited to reactions involving oxygen. For example, oxidation can occur during a reaction where an element loses electrons, such as the reaction between magnesium (Mg) and chlorine (Cl_2), where magnesium is oxidized to magnesium chloride ($MgCl_2$). Magnesium loses two electrons, increasing its oxidation state from 0 to +2 (Felton, 2012) (Figure 10.2).

Furthermore, oxidation reactions can involve the transfer of electrons between atoms or ions in redox processes. Electrons are transferred from the substance being oxidized to the substance being reduced. This transfer of electrons is essential for maintaining charge neutrality within the system. Oxidation reactions are fundamental to understanding chemical transformations and are prevalent in various fields, including organic chemistry, inorganic chemistry, electrochemistry, and biological systems.

Numerous examples illustrate oxidation reactions in various chemical systems. For instance, the combustion of hydrocarbons, such as the burning of methane (CH_4), involves the oxidation of carbon and hydrogen. In this reaction, methane combines with oxygen to produce CO_2 and water (H_2O), accompanied by the release of energy. Another example is the reaction between iron and oxygen in the presence of moisture, forming iron (III) oxide (rust). Oxidation reactions also play a vital role in the metabolism of food within living organisms, where glucose undergoes oxidation to produce CO_2 and water by releasing energy for cellular activities.

Furthermore, oxidation reactions are prevalent in the field of organic chemistry. An example is the oxidation of alcohols, where an alcohol molecule undergoes oxidation to form an aldehyde or a ketone. For instance, the oxidation of ethanol (CH_3CH_2OH) can form acetaldehyde (CH_3CHO) by losing two hydrogen atoms (Felton, 2012).

In addition, oxidation reactions are crucial in the field of electrochemistry. An example is zinc (Zn) oxidation in a zinc-carbon battery. During the discharge process, zinc atoms lose electrons and are oxidized to form zinc ions (Zn^{2+}). This oxidation process provides the necessary flow of electrons in the battery, allowing for electrical energy generation.

These examples highlight the diversified range of oxidation reactions occurring in various chemical systems, emphasizing their significance in energy production, organic synthesis, and electrochemical processes.

Understanding and manipulating oxidation reactions is essential for advancing scientific research and developing practical applications in numerous fields (Miura, 2015).

10.2.1. Oxidation State and Electron Loss

The concept of oxidation state is crucial for understanding oxidation reactions. The oxidation state, also known as oxidation number, represents the hypothetical charge that an atom would possess if the electrons in a compound were assigned according to specific rules. In an oxidation reaction, the oxidation state of an atom or ion increases due to the loss of electrons. The atom or ion that undergoes oxidation is called the reducing agent or the electron donor.

Electron loss is a key characteristic of oxidation reactions. When an atom or ion loses electrons during an oxidation process, its oxidation state becomes more positive, indicating an increase in positive charges relative to the original state. A decrease in the number of bonds with hydrogen atoms or an increase in bonds with oxygen or other electronegative elements typically accompanies this loss of electrons.

For example, in the oxidation of iron (Fe) to form iron (III) oxide (Fe_2O_3), iron loses three electrons, and its oxidation state increases from 0 to +3. This electron loss allows iron to transfer electrons to oxygen, which gains electrons and undergoes reduction.

Understanding the concept of oxidation state and the process of electron loss is vital for identifying the species undergoing oxidation in a redox reaction. It enables chemists to balance equations, determine the overall charge of compounds, and predict the behavior and reactivity of substances involved in oxidation processes (Felton, 2012).

Reduction is a chemical process characterized by the gain of electrons by a substance. It involves the decrease in the oxidation state of an atom, ion, or molecule. In a reduction reaction, a species gains electrons, which results in a more negative or less positive oxidation state. The reduction process is commonly associated with adding hydrogen atoms or removing oxygen from a compound, but it can occur without the direct involvement of hydrogen.

Numerous examples illustrate reduction reactions across various chemical systems. One example is the reaction between chlorine gas (Cl_2) and hydrogen gas (H_2) to produce hydrogen chloride (HCl). In this reaction, chlorine gains an electron from hydrogen, leading to the reduction of chlorine and the formation of HCl. Another example is the reduction of silver ions (Ag^+) to silver metal (Ag) using a reducing agent like sodium borohydride ($NaBH_4$). In this case, the silver ions gain electrons from the borohydride ions, forming silver metal (Miura, 2015).

Additionally, reduction reactions occur in various chemical systems, showcasing the diverse nature of these processes. For instance, reducing nitrate ions (NO^{3-}) to nitrogen gas (N_2) in denitrification processes by certain bacteria is an example of biological reduction. Another example is the reduction of CO_2 to form organic compounds during photosynthesis in plants, where CO_2 acts as the electron acceptor and is reduced to carbohydrates through the energy provided by sunlight.

These examples highlight the range of reduction reactions, including biological and chemical transformations, underscoring their significance in energy storage, biological processes, and environmental cycles. Understanding reduction reactions is essential for unraveling the complexities of chemical systems and exploiting their potential in various scientific and practical applications (Felton, 2012).

10.2.2. Oxidation State and Electron Gain

Similar to oxidation, the concept of oxidation state is essential in understanding reduction reactions. In a reduction process, the oxidation state of an atom or ion decreases due to the gain of electrons. The species that undergoes reduction is called the oxidizing agent or the electron acceptor. By analyzing the changes in oxidation states and the transfer of electrons, chemists can discern the occurrence of reduction reactions and their implications in chemical systems.

Electron gain is a key aspect of reduction reactions. During a reduction process, an atom or ion gains electrons, decreasing its oxidation state. An increase in the number of bonds with hydrogen atoms or a decrease in bonds with oxygen or other electronegative elements often accompanies this electron gain.

For example, each atom gains one electron in reducing chlorine gas (Cl_2) to form chloride ions (Cl^-), reducing its oxidation state from 0 to –1. This electron gain allows chlorine to accept electrons from another species, which is being oxidized (Miura, 2015).

Understanding the concept of oxidation state and the process of electron gain is crucial for identifying the species undergoing reduction in a redox reaction. It aids in balancing equations, determining the overall charge of compounds, and predicting the reactivity and behavior of substances involved in reduction processes.

10.3. BALANCING REDOX REACTIONS

10.3.1. The Need for Balancing Redox Reactions

Balancing redox reactions is essential to ensure the validity and reliability of chemical equations. Unbalanced equations violate the fundamental principles of conservation of mass and charge, leading to incorrect interpretations of the reaction and inaccurate calculations. Balancing redox reactions allows for determining precise stoichiometric ratios between reactants and products, providing a clear understanding of the number of substances involved in the reaction. This information is crucial for various applications, including chemical synthesis, environmental analysis, and industrial processes. Balanced redox equations enable scientists and engineers to make decisions and predictions about reaction outcomes, ensuring efficient and accurate experimental design and optimization (Felton, 2012).

10.3.2. The Half-Reaction Method

The half-reaction method, also known as the ion-electron method or the oxidation number method, is a systematic approach used to balance redox reactions. This method divides the redox reaction into separate oxidation and reduction half-reactions, simplifying the balancing process.

1. **Identifying the Oxidation and Reduction Half-Reactions:** Identifying the oxidation and reduction half-reactions is crucial in balancing redox reactions. For example, consider the reaction between magnesium (Mg) and oxygen (O_2) to form magnesium oxide (MgO). In this reaction, magnesium is oxidized, losing two electrons to become Mg^{2+}, while oxygen is reduced, gaining two electrons to form O^{2-}. The oxidation half-reaction is the loss of electrons by magnesium ($Mg \rightarrow Mg^{2+} + 2e^-$), and the reduction half-reaction is the gain of electrons by oxygen ($O_2 + 4e^- \rightarrow 2O^{2-}$). By identifying and balancing these half-reactions, the overall redox equation can be accurately balanced, representing the electron transfer and chemical changes occurring in the reaction. This approach applies to a wide range of redox reactions, allowing for a systematic and consistent method of balancing equations (Felton, 2012).

2. **Balancing Half-Reactions:** Balancing half-reactions is a critical step in the overall process of balancing redox equations. Adjusting the species' coefficients can conserve the number of atoms and the charge. For example, in the oxidation half-reaction of the reaction between potassium dichromate ($K_2Cr_2O_7$) and iron (II) chloride ($FeCl_2$), the chromium atoms lose electrons to form chromium (III) ions. Balancing this half-reaction involves adding 6 electrons ($6e^-$) to the left side to balance the charges. Similarly, iron (II) ions gain electrons in the reduction half-reaction to form iron (III) ions. Balancing this half-reaction involves adding 6 electrons ($6e^-$) to the right side. The overall redox equation can be accurately balanced by adjusting the coefficients and ensuring the conservation of atoms and charges in both half-reactions. This process comprehensively represents the redox reaction's electron transfer and chemical transformations (Miura, 2015).

10.3.3. Balancing the Overall Redox Reaction

Balancing the redox reaction involves combining the balanced oxidation and reduction half-reactions to conserve charge and atoms. For instance, in the reaction between potassium dichromate ($K_2Cr_2O_7$) and iron (II) chloride ($FeCl_2$), the balanced oxidation half-reaction is $6Fe^{2+} \rightarrow 6Fe^{3+} + 6e^-$, and the balanced reduction half-reaction is $Cr_2O_7^{2-} + 14H^+ + 6e^- \rightarrow 2Cr^{3+} + 7H_2O$. The number of electrons must be equal to balance the overall reaction. Multiplying the oxidation half-reaction by six and the reduction half-reaction by one allows the 6 electrons to cancel out, resulting in a balanced overall redox equation:

$$6Fe^{2+} + Cr_2O_7^{2-} + 14H^+ \rightarrow 6Fe^{3+} + 2Cr^{3+} + 7H_2O$$

By achieving balance at both the atom and charge levels, the redox equation accurately represents the electron transfer and chemical changes occurring in the reaction.

10.3.4. Balancing Redox Reactions in Acidic Solutions

Balancing redox reactions in acidic solutions requires the addition of hydrogen ions (H^+) and water molecules (H_2O) to the half-reactions. These additions help balance the charges and ensure the conservation of mass (Felton, 2012).

The steps for the balancing redox reactions in acidic solutions involve the following:

- Balancing the atoms other than hydrogen and oxygen.
- Balancing oxygen atoms by adding water molecules.
- Balancing hydrogen atoms by adding hydrogen ions.
- Balancing the charges by adding electrons.
- Multiplying the half-reactions to ensure an equal number of electrons in both.

An example is the reaction between potassium dichromate ($K_2Cr_2O_7$) and iron (II) chloride ($FeCl_2$) to form chromium (III) chloride ($CrCl_3$), iron (III) chloride ($FeCl_3$), and potassium chloride (KCl). Applying the above steps balances the redox equation in an acidic solution (Miura, 2015).

10.3.5. Balancing Redox Reactions in Basic Solutions

Balancing redox reactions in basic solutions requires additional steps to account for the presence of hydroxide ions (OH^-) and water molecules.

The steps for balancing redox reactions in basic solutions involve the following:

- Balancing the atoms other than hydrogen and oxygen.
- Balancing oxygen atoms by adding water molecules.
- Balancing hydrogen atoms by adding hydrogen ions (in acidic solution) and hydroxide ions (in basic solution).
- Balancing the charges by adding electrons.
- Adjust the number of water molecules to ensure equal hydrogen and hydroxide ions.

An example is the reaction between potassium permanganate ($KMnO_4$) and sodium sulfite (Na_2SO_3) to form manganese dioxide (MnO_2), sodium sulfate (Na_2SO_4), and potassium hydroxide (KOH). The redox equation can be balanced in a basic solution by applying the abovementioned steps (Felton, 2012).

10.4. ELECTROCHEMICAL CELLS

Electrochemical cells convert chemical energy into electrical energy through redox reactions. They consist of two electrodes, an electrolyte, and an external circuit. The electrodes are immersed in an electrolyte solution containing ions participating in redox reactions. Electrons flow through the external circuit when a wire connects the electrodes, generating an electric current. Electrochemical cells are widely used in various applications, including batteries, fuel cells, and electrolysis processes.

10.4.1. The Daniel Cell

The Daniel cell is renowned for its stable and reliable operation, making it an important milestone in developing electrochemical cells. It served as the foundation for understanding the principles of electrochemistry and provided valuable insights into the concept of cell potential. Using zinc and copper electrodes in the Daniel cell allows for efficient electron transfer and is a model system for studying redox reactions. Furthermore, separating the two electrolyte solutions by a porous barrier minimizes mixing and facilitates controlled ion flow, ensuring the cell's longevity. The Daniel cell remains a valuable tool in educational settings and continues to be studied for its historical significance and fundamental contributions to electrochemistry (Miura, 2015).

The Daniel cell is an early example of an electrochemical cell invented by John Frederic Daniel in 1836. It consists of a zinc electrode immersed in a zinc sulfate solution and a copper electrode immersed in a solution of copper sulfate. The two electrolyte solutions are separated by a porous barrier to prevent direct mixing. The Daniel cell operates based on the redox reaction between zinc and copper, where zinc undergoes oxidation at the anode ($Zn \rightarrow Zn^{2+} + 2e^-$), and copper ions are reduced at the cathode ($Cu^{2+} + 2e^- \rightarrow Cu$).

10.4.2. The Galvanic Cell

The galvanic cell, also known as a voltaic cell, is a type of electrochemical cell that generates electrical energy from spontaneous redox reactions. It consists of two half-cells, each with an electrode and an electrolyte. The half-cells are connected by a salt bridge or a porous membrane, allowing ion flow. The redox reaction occurs spontaneously in the galvanic cell, producing an electric current. The most common example is the zinc-copper galvanic cell.

10.4.3. Galvanic Cell

The galvanic cell has widespread applications in various fields, including batteries, power generation, and electrochemical sensors. It harnesses the energy released from spontaneous redox reactions to power electronic devices. Using two half-cells, each with its electrode and electrolyte, allows for efficient separation of the reactants and products, ensuring a controlled flow of ions. The salt bridge or porous membrane facilitates the exchange of ions, maintaining charge neutrality and preventing the mixing of electrolytes. The galvanic cell's ability to generate a continuous electric current is a key component in portable electronic devices and renewable energy systems. Its simplicity and versatility makes it an ideal system for studying electrochemical processes and investigating fundamental principles of thermodynamics and chemical kinetics (Felton, 2012).

10.4.4. The Electrolytic Cell

Unlike galvanic cells, electrolytic cells require an external power source to drive non-spontaneous redox reactions. They are used for electrolysis processes, decomposing chemical compounds into constituent elements or ions. An electrolytic cell consists of two electrodes, the anode (positive) and cathode (negative), immersed in an electrolyte solution. When an electric current is applied, oxidation occurs at the anode, and reduction occurs at the cathode (Miura, 2015).

The electrolytic cell is a powerful tool for various industrial processes, such as electroplating, metal extraction, and water electrolysis. Its ability to drive non-spontaneous redox reactions using an external power source enables the manipulation and transformation of chemical compounds on a large scale. The electrolytic cell allows precise control over the desired chemical reactions by controlling the magnitude and direction of the applied electric current. The anode, acting as the positive electrode, attracts negatively charged ions, inducing their oxidation. Conversely, the cathode, functioning as the negative electrode, attracts positively charged ions, promoting their reduction. The electrolytic cell's versatility and controllability makes it an indispensable asset in industries requiring precise chemical transformations and producing specific materials with tailored properties (Felton, 2012).

10.4.5. Cell Potential and the Nernst Equation

Cell potential, also known as electromotive force (EMF), measures the driving force of an electrochemical cell. It represents the potential difference between the two electrodes and is measured in volts (V). Cell potential determines whether a redox reaction is spontaneous or non-spontaneous. A positive cell potential indicates a spontaneous reaction, while a negative one indicates a non-spontaneous reaction.

1. **The Nernst Equation and its Application:** The Nernst equation

is a mathematical relationship that relates the cell potential to the concentrations of reactants and products involved in the electrochemical reaction. It allows for calculating cell potential under non-standard conditions, considering the activities or concentrations of species present. The Nernst equation is expressed as:

$$E = E° - (RT/nF) \times \ln(Q)$$

E is the cell potential, E° is the standard cell potential, R is the gas constant, T is the temperature in Kelvin, n is the number of electrons transferred, F is the Faraday constant, and Q represents the reaction quotient (Miura, 2015).

The Nernst equation is essential for predicting and understanding the behavior of electrochemical cells under various conditions. It enables the determination of equilibrium constants, the effect of concentration changes on cell potential, and the feasibility of redox reactions. By utilizing the Nernst equation, scientists and engineers can optimize the design and operation of electrochemical systems, such as fuel cells, batteries, and corrosion prevention strategies.

For example, in a concentration cell, where the same species are present but at different concentrations in the two half-cells, the Nernst equation allows calculating the cell potential based on the species concentrations. This information can be used to determine the direction and rate of the chemical reaction taking place in the cell. Additionally, the Nernst equation is used in pH measurements using glass electrodes, as the potential difference across the electrode is proportional to the logarithm of the hydrogen ion concentration (Felton, 2012) (Figure 10.3).

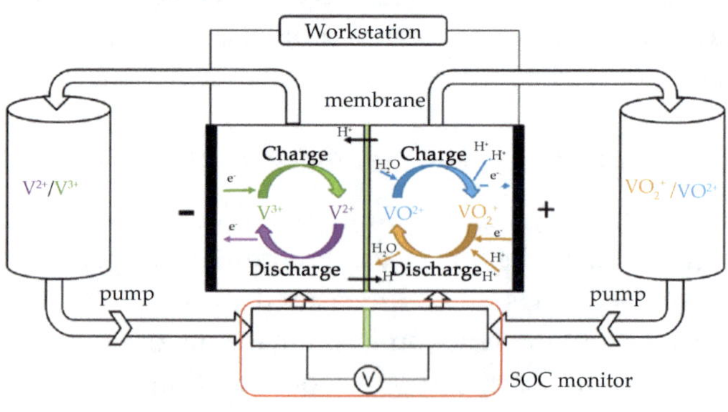

Figure 10.3. *Vanadium redox flow battery.*

Source: Yin, S., Zhou, L., Du, X., et al., (2019). Influence of temperature on performance of all vanadium redox flow battery: Analysis of ionic mass transfer. Ionics, 25, 593–606. https://doi.org/10.1007/s11581-018-2626-z.

10.5. REDOX REACTIONS IN BIOLOGICAL SYSTEMS

10.5.1. Redox Reactions in Cellular Respiration

1. **Redox Reactions are Involved in Glycolysis:** Cellular respiration is the metabolic process of cells obtaining energy from organic molecules, such as glucose, through redox reactions. It involves three main stages: glycolysis, the citric acid cycle (or Krebs cycle), and the electron transport chain. Redox reactions play a crucial role in each stage, facilitating the transfer of electrons and the generation of ATP (Miura, 2015).
2. **Redox Reactions in Glycolysis:** Glycolysis is the initial step of cellular respiration, occurs in the cytoplasm. It involves the breakdown of glucose into two molecules of pyruvate. Several redox reactions occur throughout glycolysis, resulting in glucose oxidation and the reduction of electron carriers like NAD^+ to NADH.

   In glycolysis, the redox reactions occur in two key steps. The first redox reaction involves the oxidation of glucose to form glyceraldehyde-3-phosphate, accompanied by the reduction of NAD^+ to NADH. The enzyme glyceraldehyde-3-phosphate dehydrogenase catalyzes this reaction. The second redox reaction occurs during the conversion of 1,3-bisphosphoglycerate to 3-phosphoglycerate, where NADH is again generated as NAD^+ is reduced (Felton, 2012).

   These redox reactions in glycolysis serve multiple purposes. They generate NADH, a high-energy electron carrier, and produce ATP through substrate-level phosphorylation. Furthermore, the oxidation of glucose to pyruvate releases energy that can be utilized by the cell.

   Glycolysis is a central metabolic pathway found in both aerobic and anaerobic organisms. It provides a vital source of energy and building blocks for cellular processes. By understanding the redox reactions within glycolysis, scientists can gain insights into the regulation and coordination of metabolic pathways and their significance in various biological systems.

3. **Redox Reactions in the Citric Acid Cycle:** The citric acid cycle occurs in the mitochondria and completes the breakdown of glucose. It involves a series of redox reactions that transfer high-energy electrons to carrier molecules, such as NAD^+ and FAD, forming NADH and $FADH_2$. These electron carriers then donate electrons to the electron transport chain.

   The redox reactions in the citric acid cycle occur at several key steps. For example, during the conversion of isocitrate to α-ketoglutarate, isocitrate is oxidized, leading to the reduction of NAD^+ to NADH. Similarly, in the conversion of α-ketoglutarate to succinyl-CoA, α-ketoglutarate is oxidized, generating NADH (Miura, 2015).

   These redox reactions serve multiple purposes within the citric acid cycle. They produce high-energy electron carriers, NADH and $FADH_2$, and generate ATP through substrate-level phosphorylation. Additionally, the cycle replenishes intermediates for other metabolic pathways and provides precursors for synthesizing molecules such as amino acids.

   The citric acid cycle is a central hub for energy metabolism, linking various metabolic pathways and playing a critical role in the overall energy production of the cell. Understanding the redox reactions within the citric acid cycle contributes to our knowledge of cellular respiration and the efficient utilization of energy sources in biological systems.

4. **Redox Reactions in the Electron Transport Chain:** The electron transport chain is located in the inner mitochondrial membrane. It consists of a series of protein complexes and electron carriers that facilitate the transfer of electrons. As electrons pass through the chain, energy is released to pump protons across the membrane, establishing an electrochemical gradient. This gradient is then used to synthesize ATP through oxidative phosphorylation (Felton, 2012).

   A series of redox reactions drive the electron transport chain. As electrons are transferred from electron carriers, such as NADH and $FADH_2$, to protein complexes within the chain, the carriers are oxidized back to NAD^+ and FAD. The final electron acceptor in the chain is oxygen, which serves as the ultimate oxidizing agent.

The redox reactions in the electron transport chain play a crucial role in ATP production. As electrons flow through the chain, the energy released pumps protons across the mitochondrial membrane. This creates an electrochemical gradient with a higher concentration of protons in the intermembrane space. The flow of protons back into the mitochondrial matrix through ATP synthase drives the synthesis of ATP.

The electron transport chain is a highly efficient process that generates the majority of ATP in aerobic organisms. It demonstrates the intricate coordination of redox reactions and the coupling of electron transfer with proton pumping to produce ATP, the universal energy currency of cells. Understanding the mechanisms and regulation of these redox reactions provides insights into cellular energy metabolism and its significance in biological systems (Miura, 2015).

10.5.2. Redox Reactions in Photosynthesis

Photosynthesis is the process by which plants, algae, and some bacteria convert sunlight into chemical energy in the form of glucose. It involves two main stages: the light-dependent reactions and the Calvin cycle. Redox reactions play a vital role in both stages (Felton, 2012).

1. **Redox Reactions in the Light-Dependent Reactions:** The light-dependent reactions occur in the thylakoid membranes of chloroplasts. These reactions capture light energy and convert it into chemical energy through ATP and NADPH. Chlorophyll and other pigments absorb light, exciting electrons and initiating a series of redox reactions that generate ATP and reduce $NADP^+$ to NADPH.

   These reactions are essential for capturing and utilizing light energy. Chlorophyll and other pigments in the thylakoid membranes of chloroplasts absorb photons, causing the excitation of electrons. This excitation triggers a chain of redox reactions, where electrons are transferred from the excited chlorophyll molecules to electron carriers.

   As electrons move through the electron transport chain in the thylakoid membrane, energy is released and used to generate ATP through chemiosmosis. The flow of electrons also leads to the reduction of $NADP^+$ to NADPH, which serves as a high-energy electron carrier in the subsequent dark reactions of photosynthesis.

The light-dependent redox reactions are crucial for converting light energy into chemical energy and providing the necessary reducing power (NADPH) and ATP for synthesizing sugars and other organic molecules. These reactions exemplify the intricacies of electron transfer and the utilization of redox chemistry in harnessing light energy for the sustenance of life.

2. **Redox Reactions in the Calvin Cycle:** The Calvin cycle takes place in the stroma of chloroplasts. It utilizes ATP and NADPH produced in light-dependent reactions to convert CO_2 into glucose. The cycle involves a series of redox reactions that use ATP and NADPH as energy sources to fix carbon and reduce it to form glucose.

   Redox reactions are vital in converting CO_2 into glucose in the Calvin cycle. The cycle utilizes the energy-rich molecules ATP and NADPH, produced in the light-dependent reactions, as energy sources and reducing power. Through a series of enzymatic reactions, CO_2 is fixed and progressively reduced, resulting in the formation of glucose (Miura, 2015).

The redox reactions in the Calvin cycle involve the transfer of high-energy electrons from NADPH to the CO_2 molecules, leading to their reduction and the synthesis of glucose. ATP provides the necessary energy to drive these reactions, ensuring the incorporation of carbon and the formation of complex organic molecules.

By relying on redox chemistry, the Calvin cycle enables plants and other photosynthetic organisms to utilize the energy captured from sunlight to convert inorganic CO_2 into energy-rich carbohydrates. These reactions highlight the intricate interplay between redox reactions, energy transfer, and carbon fixation in photosynthesis (Felton, 2012).

10.5.3. Redox Reactions in Antioxidants

Antioxidants can inhibit or prevent oxidation by donating electrons or neutralizing free radicals. They play a crucial role in maintaining redox homeostasis and protecting cells from oxidative damage caused by reactive oxygen species (ROS).

1. **Redox Reactions and Oxidative Stress:** Oxidative stress

occurs when there is an imbalance between ROS production and antioxidants' ability to neutralize them. ROS, such as superoxide radicals and hydrogen peroxide, are highly reactive and can damage cellular components. Redox reactions involving antioxidants help to mitigate oxidative stress by neutralizing ROS and maintaining cellular redox balance.

Antioxidants act as scavengers, donating electrons or hydrogen atoms to neutralize ROS. This process helps to prevent ROS's harmful effects on cellular components, including lipids, proteins, and DNA. By maintaining a balanced redox state, antioxidants protect cells from oxidative damage and promote overall cellular health. The intricate interplay between redox reactions and antioxidants is essential in maintaining redox homeostasis and mitigating the detrimental effects of oxidative stress in biological systems (Miura, 2015).

2. **Role of Antioxidants in Redox Homeostasis:** The role of antioxidants in redox homeostasis is crucial in maintaining cellular health. Antioxidants effectively neutralize reactive species and prevent oxidative damage by participating in redox reactions. Enzymatic antioxidants, such as superoxide dismutase and catalase, convert harmful reactive species into less reactive forms. Non-enzymatic antioxidants, such as vitamin C, E, and glutathione, directly donate electrons to reactive species, effectively reducing their reactivity. Together, these antioxidants work in harmony to maintain the delicate balance of redox reactions and protect cells from oxidative stress, ensuring the proper functioning of biological systems (Felton, 2012).

In addition to their direct involvement in redox reactions, antioxidants play several other important roles in redox homeostasis. They help regulate signaling pathways and gene expression related to cellular stress responses. Antioxidants also can repair oxidative damage to biomolecules such as DNA, proteins, and lipids. By preventing excessive oxidative damage, antioxidants help preserve cellular components' integrity and maintain their proper function.

Moreover, antioxidants have been linked to various health benefits. Their ability to counteract oxidative stress has been associated with reduced risk of chronic diseases such as cardiovascular diseases, neurodegenerative disorders, and certain types of cancer. Antioxidants also contribute to immune system function by protecting immune cells from oxidative damage. It is worth to note that maintaining an appropriate balance of antioxidants is crucial. While an inadequate antioxidant defense system can lead to oxidative stress and cellular damage, an excess of antioxidants can interfere with normal redox signaling processes, potentially affecting cellular function. Therefore, achieving a balanced and diverse antioxidant intake through a healthy diet rich in fruits, vegetables, and other antioxidant-rich foods is important for overall health and well-being (Miura, 2015).

10.6. INDUSTRIAL APPLICATIONS OF REDOX REACTIONS

10.6.1. Redox Reactions in Corrosion

Corrosion refers to the gradual degradation of metals due to chemical reactions with the environment. It involves the oxidation of metal atoms, forming metal oxides, hydroxides, or other compounds. Corrosion is a significant concern in various industries, leading to structural damage, decreased durability, and economic losses.

1. **Types of Corrosion Reactions:** There are different types of corrosion reactions, including galvanic corrosion, pitting corrosion, crevice corrosion, and stress corrosion cracking. Galvanic corrosion occurs when two dissimilar metals are in contact with an electrolyte, creating a potential difference and initiating a redox reaction. Pitting corrosion involves localized damage in pits or holes on the metal surface. Crevice corrosion occurs in narrow spaces where the metal is exposed to stagnant electrolytes. Stress corrosion cracking happens when tensile stress and a corrosive environment leads to crack formation and propagation (Felton, 2012).

   Another type of corrosion reaction is intergranular corrosion, which occurs along the grain boundaries of a metal. It is caused by the preferential attack of the grain boundaries due to differences in chemical composition or impurities. Uniform corrosion is another common type, where the metal corrodes evenly over its entire surface. Environmental factors such as temperature, humidity, pH, and corrosive substances can influence the occurrence and severity of these corrosion reactions. Understanding and effectively managing these various corrosion reactions are crucial in industrial settings to ensure the longevity and reliability of metal structures and components.

2. **Prevention and Control of Corrosion:** Several methods are employed to prevent and control corrosion. Protective coatings, such as paints, varnishes, and metal plating, act as barriers, preventing direct contact between the metal and corrosive agents. Cathodic protection involves the application of a sacrificial anode or an impressed current to protect the metal from corrosion. Proper material selection, design considerations, and regular inspection and maintenance are crucial in corrosion prevention and control (Miura, 2015).

10.6.2. Redox Reactions in Batteries

Batteries are energy storage devices that rely on redox reactions to generate and deliver electrical energy. They consist of one or more electrochemical cells, each containing an anode (negative electrode) and a cathode (positive electrode), separated by an electrolyte. Batteries are widely used in various applications, including portable electronics, electric vehicles, and grid energy storage.

10.6.2.1. Redox Reactions in Different Types of Batteries

Different types of batteries employ specific redox reactions. For example, in lead-acid batteries, the redox reaction occurs between lead dioxide (PbO_2) at the positive electrode and lead (Pb) at the negative electrode. In lithium-ion batteries, the redox reactions involve the movement of lithium ions between the positive and negative electrodes. Other types of batteries, such as nickel-cadmium (Ni-Cd) and nickel-metal hydride (NiMH) batteries, also involve redox reactions of different elements or compounds (Felton, 2012).

10.6.3. Redox Reactions in Electroplating

Electroplating is a process that involves the deposition of a metal layer onto a substrate through the use of an electric current. It is widely used in various industries for decorative purposes, corrosion protection, and enhancement of mechanical properties. Electroplating relies on redox reactions to transfer metal ions from an electrolyte solution onto the substrate.

10.6.4. Redox Reactions in the Electroplating Process

In electroplating, the substrate to be plated is connected to the cathode, and a metal electrode is connected to the anode. The electrolyte solution contains metal ions of the desired plating material. When an electric current is applied, the metal ions from the electrolyte undergo reduction at the cathode, forming a metal layer on the substrate. Simultaneously, the metal electrode at the anode oxidizes, releasing metal ions into the electrolyte to maintain the ion concentration. The redox reactions involved in electroplating ensure the controlled deposition of the desired metal layer with specific properties onto the substrate (Miura, 2015).

10.7. ENVIRONMENTAL SIGNIFICANCE OF REDOX REACTIONS

10.7.1. Redox Reactions in Environmental Processes

1. **Redox Reactions in the Nitrogen Cycle:** The nitrogen cycle is a fundamental environmental process, and redox reactions are crucial in its various stages. Nitrogen exists in different forms, including atmospheric nitrogen (N_2), ammonia (NH_3), nitrite (NO^{2-}), and nitrate (NO^{3-}). Redox reactions mediated by bacteria are responsible for converting nitrogen between these forms. Nitrogen-fixing bacteria, such as Rhizobium, converts atmospheric nitrogen into ammonia through a reduction process called nitrogen fixation. Nitrifying bacteria, like Nitrosomonas and Nitrobacter, perform oxidation reactions to convert ammonia to nitrite and nitrate. Denitrifying bacteria, such as Pseudomonas, reduce nitrate back to atmospheric nitrogen, completing the cycle. These redox reactions ensure nitrogen availability for plants and other organisms while maintaining a balance in nitrogen levels in ecosystems (Miura, 2015).
2. **Redox Reactions in the Carbon Cycle:** The carbon cycle is another critical environmental process that involves redox reactions. Through photosynthesis, green plants and algae convert CO_2 from the atmosphere into organic compounds like glucose. This reduction reaction utilizes light energy and involves the transfer of electrons from water to CO_2. During respiration and decomposition, organic compounds oxidize, releasing CO_2 into the atmosphere. These redox reactions involving carbon ensure

the circulation and balance of carbon in ecosystems. Carbon sequestration processes, such as converting CO_2 into organic matter by marine organisms or storing it in geological formations, also involve redox reactions and significantly mitigate climate change also (Felton, 2012).

10.7.2. Redox Reactions and Water Treatment

1. **Redox Reactions in Water Purification:** Water purification processes utilize redox reactions to remove contaminants and ensure the safety of drinking water. Oxidation-reduction reactions eliminate harmful pollutants, including organic compounds and microorganisms. For example, in disinfection, strong oxidizing agents like chlorine or ozone destroy pathogenic microorganisms through redox reactions. The oxidizing agent donates electrons, causing the oxidation of cellular components in the microorganisms, leading to their inactivation or death. This ensures that waterborne diseases are prevented and water quality is maintained.
2. **Application of Redox Reactions in Wastewater Treatment:** Redox reactions are also extensively utilized in wastewater treatment to remove pollutants and protect the environment. Chemical precipitation and coagulation processes involve adding chemicals that undergo redox reactions with pollutants, forming insoluble compounds or flocs that can be easily removed. Redox reactions occur at the electrodes in electrochemical treatments, where oxidation or reduction processes occur to degrade and remove contaminants. Electrochemical oxidation can break down complex organic compounds into simpler forms, while electrochemical reduction can convert harmful ions into less toxic forms. These applications of redox reactions in wastewater treatment help to ensure the removal of pollutants and the protection of water resources and ecosystems (Miura, 2015).

10.8. SUMMARY OF KEY POINTS

We have delved into their various aspects and applications in exploring redox reactions. Redox reactions involve the transfer of electrons between species, leading to oxidation and reduction processes. We discussed their importance in numerous fields, including chemistry, biology, industry, and the environment. We explored redox reactions in chemical systems, where they are used for energy production, synthesis of compounds, and as a basis for analytical techniques. Understanding redox reactions is crucial for balancing chemical equations and predicting the behavior of reactions (Felton, 2012).

Redox reactions play a fundamental role in cellular respiration, photosynthesis, and antioxidant defense mechanisms in biological systems. They enable the conversion of energy, the synthesis of biomolecules, and the maintenance of cellular redox balance.

The industrial applications of redox reactions are diverse and significant. We explored their involvement in corrosion processes, batteries, and electroplating, highlighting their importance in everyday technologies and materials.

Redox reactions also have profound environmental significance. We discussed their role in the nitrogen and carbon cycles, which are critical for the functioning of ecosystems and the balance of atmospheric gases. Furthermore, redox reactions find applications in water treatment, ensuring drinking water purification and removing pollutants from wastewater (Miura, 2015).

The significance of redox reactions is vast and far-reaching. They are at the core of energy conversion processes, chemical synthesis, and the functioning of biological systems. Redox reactions have enabled technological advancements and are pivotal in various industrial applications.

Understanding redox reactions allows us to comprehend natural processes, such as the cycling of elements in ecosystems and the maintenance of environmental balance. Redox reactions play a vital role in the sustainability of our planet and the preservation of its resources (Miura, 2015).

Redox reactions are significant in medicine, renewable energy, agriculture, and material development. They are involved in drug metabolism, energy conversion, nutrient availability, and environmental remediation. Redox reactions shape our world, addressing challenges and promoting sustainability. Continued research in redox chemistry promises innovative solutions for a brighter future.

The field of redox chemistry continues to evolve, offering exciting prospects for the future. With the increasing demand for sustainable energy sources, researchers are exploring innovative ways to harness and store energy using redox reactions. Advancements in battery technologies, fuel cells, and solar energy systems rely on understanding and optimizing redox processes (Miura, 2015).

Moreover, developing green chemistry principles aims to minimize the environmental impact of chemical processes. Redox chemistry plays a crucial role in these efforts, enabling the design of sustainable and eco-friendly processes for synthesis and waste treatment (Felton, 2012).

Further research into redox reactions will deepen our understanding of biological systems, unveiling new insights into diseases, metabolism, and cellular functions. This knowledge can pave the way for novel therapeutic strategies and medical advancements.

Redox reactions are of immense importance in diverse fields. Redox reactions offer countless opportunities for innovation, sustainability, and improving our understanding of the world around us. Embracing the potential of redox chemistry will undoubtedly shape the future of science, technology, and the environment (Felton, 2012).

Bibliography

1. Anderson, D. F., & Kurtz, T. G., (2011). Continuous time Markov chain models for chemical reaction networks. In: *Design and Analysis of Biomolecular Circuits: Engineering Approaches to Systems and Synthetic Biology* (pp. 3–42). New York, NY: Springer New York.
2. Anscher, D., (2011). Double replacement reactions. *Journal of the Colorado-Wyoming Academy of Science*, *33*(1), 34–35.
3. Bédard, A. C., Adamo, A., Aroh, K. C., Russell, M. G., Bedermann, A. A., Torosian, J., & Jamison, T. F., (2018). Reconfigurable system for automated optimization of diverse chemical reactions. *Science*, *361*(6408), 1220–1225.
4. Chen, L., Kuai, L., Yu, X., Li, W., & Geng, B., (2013). Advanced catalytic performance of Au–Pt double-walled nanotubes and their fabrication through galvanic replacement reaction. *Chemistry–A European Journal*, *19*(35), 11753–11758.
5. Collins, K. D., & Glorius, F., (2013). A robustness screen for the rapid assessment of chemical reactions. *Nature Chemistry*, *5*(7), 597–601.
6. Collins, M. N., & Birkinshaw, C., (2013). Hyaluronic acid-based scaffolds for tissue engineering—A review. *Carbohydrate Polymers*, *92*(2), 1262–1279.
7. Davis, M. E., & Davis, R. J., (2012). *Fundamentals of Chemical Reaction Engineering*. Courier Corporation.
8. De Andrade, M. R. (2020). *A Priori Components of Science: Lavoisier and the Law of Conservation of Mass in Chemical Reactions*, *2*(1), 14–23.

9. Farhodovna, A. M., & Abdishukurovna, E. M., (2021). Application of the law of conservation of energy in economics. *The American Journal of Applied Sciences*, *3*(01), 93–103.
10. Felton, R. H., (2012). Primary redox reactions of metalloporphyrins. *The Porphyrins*, *5*, 53–115.
11. Fernández, I., & Bickelhaupt, F. M., (2014). The activation strain model and molecular orbital theory: understanding and designing chemical reactions. *Chemical Society Reviews*, *43*(14), 4953–4967.
12. Fleming, I., (2011). *Molecular Orbitals and Organic Chemical Reactions*. John Wiley & Sons.
13. Georgievskii, Y., Miller, J. A., Burke, M. P., & Klippenstein, S. J., (2013). Reformulation and solution of the master equation for multiple-well chemical reactions. *The Journal of Physical Chemistry A*, *117*(46), 12146–12154.
14. Glassman, I., Yetter, R. A., & Glumac, N. G., (2014). *Combustion*. Academic press.
15. Greening, C., Ahmed, F. H., Mohamed, A. E., Lee, B. M., Pandey, G., Warden, A. C., & Jackson, C. J., (2016). Physiology, biochemistry, and applications of F420-and Fo-dependent redox reactions. *Microbiology and Molecular Biology Reviews*, *80*(2), 451–493.
16. Grondal, C., Jeanty, M., & Enders, D., (2010). Organocatalytic cascade reactions as a new tool in total synthesis. *Nature Chemistry*, *2*(3), 167–178.
17. Grossmann, A., & Enders, D., (2012). N-heterocyclic carbene catalyzed domino reactions. *Angewandte Chemie International Edition*, *51*(2), 314–325.
18. Hansgen, D. A., Vlachos, D. G., & Chen, J. G., (2010). Using first principles to predict bimetallic catalysts for the ammonia decomposition reaction. *Nature Chemistry*, *2*(6), 484–489.
19. Kącka-Zych, A., Ríos-Gutiérrez, M., & Domingo, L. R., (2019). A molecular electron density theory study of the Lewis acid–catalyzed decomposition reaction of nitroethyl benzoate using aluminum derivatives. *Journal of Physical Organic Chemistry*, *32*(6), e3938.
20. Kim, M. H., Lu, X., Wiley, B., Lee, E. P., & Xia, Y., (2010). Morphological evolution of single-crystal Ag nanospheres during the galvanic replacement reaction with HAuCl4. *The Journal of Physical Chemistry C*, *112*(21), 7872–7876.
21. Kohen, R., & Nyska, A., (2012). Invited review: oxidation of biological systems: oxidative stress phenomena, antioxidants, redox reactions, and methods for their quantification. *Toxicologic Pathology*, *30*(6), 620–650.

22. Lee, J. W., Kim, H. S., & Park, N. G., (2016). Lewis acid–base adduct approach for high efficiency perovskite solar cells. *Accounts of Chemical Research*, *49*(2), 311–319.

23. Li, Q., Xia, W., Yao, L., & Shao, Y., (2012). Anharmonic effect of the decomposition reaction of the CF3CCl2O radical. *Canadian Journal of Chemistry*, *90*(2), 186–194.

24. Liao, L., Wang, S., Xiao, J., Bian, X., Zhang, Y., Scanlon, M. D., & Girault, H. H., (2014). A nanoporous molybdenum carbide nanowire as an electrocatalyst for hydrogen evolution reaction. *Energy & Environmental Science*, *7*(1), 387–392.

25. Merki, D., & Hu, X., (2011). Recent developments of molybdenum and tungsten sulfides as hydrogen evolution catalysts. *Energy & Environmental Science*, *4*(10), 3878–3888.

26. Metiu, H., Chrétien, S., Hu, Z., Li, B., & Sun, X., (2012). Chemistry of Lewis acid–base pairs on oxide surfaces. *The Journal of Physical Chemistry C*, *116*(19), 10439–10450.

27. Obaya, A. V., Barocio, Y. R., & Rodríguez, Y. M. V., (2021). Online Simulators for the Teaching of the Law of Conservation of Matter and Chemical Reactions in High School. *Science Education International*, *32*(3), 209–219.

28. Oh, M. H., Yu, T., Yu, S. H., Lim, B., Ko, K. T., Willinger, M. G., & Hyeon, T., (2013). Galvanic replacement reactions in metal oxide nanocrystals. *Science*, *340*(6135), 964–968.

29. Oriji, G., Katayama, Y., & Miura, T., (2015). Investigations on V (IV)/V (V) and V (II)/V (III) redox reactions by various electrochemical methods. *Journal of Power Sources*, *139*(1–2), 321–324.

30. Parimalam, B. S., MacIntosh, A. D., Kadam, R., & Lucht, B. L., (2017). Decomposition reactions of anode solid electrolyte interphase (SEI) components with LiPF6. *The Journal of Physical Chemistry C*, *121*(41), 22733–22738.

31. Parker, J. M., de los Santos, E. X., & Anderson, C. W., (2013). What learning progressions on carbon-transforming processes tell us about how students learn to use the laws of conservation of matter and energy. *Educación química*, *24*(4), 399–406.

32. Prentice, B. M., & McLuckey, S. A., (2013). Gas-phase ion/ion reactions of peptides and proteins: acid/base, redox, and covalent chemistries. *Chemical Communications*, *49*(10), 947–965.

33. Putnis, A., (2010). Mineral replacement reactions. *Reviews in Mineralogy and Geochemistry*, *70*(1), 87–124.

34. Rogachev, A. S., & Mukasyan, A. S., (2014). *Combustion for Material Synthesis*. CRC press.
35. Sanders, W. C., Ainsworth, P. D., Archer Jr, D. M., Armajo, M. L., Emerson, C. E., Calara, J. V., & Swenson, J. D., (2014). Characterization of micro-and nanoscale silver wires synthesized using a single-replacement reaction between sputtered copper metal and dilute silver nitrate solutions. *Journal of Chemical Education*, *91*(5), 705–710.
36. Sarathy, S. M., Oßwald, P., Hansen, N., & Kohse-Höinghaus, K., (2014). Alcohol combustion chemistry. *Progress in energy and Combustion Science*, *44*, 40–102.
37. Schwaller, P., Gaudin, T., Lanyi, D., Bekas, C., & Laino, T., (2018). "Found in Translation": predicting outcomes of complex organic chemistry reactions using neural sequence-to-sequence models. *Chemical Science*, *9*(28), 6091–6098.
38. Schwartz, W. R., & Pfefferle, L. D., (2012). Combustion of methane over palladium-based catalysts: support interactions. *The Journal of Physical Chemistry C*, *116*(15), 8571–8578.
39. Singh, S., Bothara, S. B., Singh, S., Patel, R., & Ughreja, R., (2011). Preliminary Pharmaceutical Characterization of some Flowers as natural indicator: acid-Base titration. *Pharmacognosy Journal*, *3*(22), 39–43.
40. Thompson, S., & Lotter, C., (2014). Conservation of matter in the life sciences. *Science Scope*, *38*(2), 57.
41. Tillet, G., Boutevin, B., & Ameduri, B., (2011). Chemical reactions of polymer crosslinking and post-crosslinking at room and medium temperature. *Progress in Polymer Science*, *36*(2), 191–217.
42. Varma, A., Mukasyan, A. S., Rogachev, A. S., & Manukyan, K. V., (2016). Solution combustion synthesis of nanoscale materials. *Chemical Reviews*, *116*(23), 14493–14586.
43. Wei, J. N., Duvenaud, D., & Aspuru-Guzik, A., (2016). Neural networks for the prediction of organic chemistry reactions. *ACS Central Science*, *2*(10), 725–732.
44. Winstein, S., & Adams, R., (2014). The Role of Neighboring Groups in Replacement Reactions. XIV. The 5, 6-Double Bond in Cholesteryl p-Toluenesulfonate as a Neighboring Group. *Journal of the American Chemical Society*, *70*(2), 838–840.
45. Winterbourn, C. C., & Kettle, A. J., (2013). Redox reactions and microbial killing in the neutrophil phagosome. *Antioxidants & Redox Signaling*, *18*(6), 642–660.

46. Zhang, C., Yang, X., Zhang, Z., Li, J., Yi, J., & Ma, H., (2022). The effect of oxidation of Al (1 1 1) surface on the catalytic efficiency of decomposition of ammonium perchlorate: A density functional theory study. *Applied Surface Science*, *602*, 154381.
47. Zhang, H., Jin, M., Wang, J., Li, W., Camargo, P. H., Kim, M. J., & Xia, Y., (2011). Synthesis of Pd– Pt bimetallic nanocrystals with a concave structure through a bromide-induced galvanic replacement reaction. *Journal of the American Chemical Society*, *133*(15), 6078–6089.
48. Zhang, J., Yu, R., Li, J., Zhai, H., Tan, G., & Tang, X., (2022). Transformation of undesired Li2CO3 into lithiophilic layer via double replacement reaction for garnet electrolyte engineering. *Energy & Environmental Materials*, *5*(3), 962–968.
49. Zheng, Y., Jiao, Y., Jaroniec, M., & Qiao, S. Z., (2015). Advancing the electrochemistry of the hydrogen-evolution reaction through combining experiment and theory. *Angewandte Chemie International Edition*, *54*(1), 52–65.

INDEX

L

M

N

O

P

Q

R

S

T

U

Printed in the United States
by Baker & Taylor Publisher Services